SURVEY OF CARPENTRY MASTERPIECES

The Modern Carpenter Joiner and Cabinet-Maker

SURVEY OF CARPENTRY MASTERPIECES

G. Lister Sutcliffe, Editor
Associate of the Royal Institute of British Architects, member of the Sanitary Institute, editor and joint-author of *Modern House Construction,* author of *Concrete: Its Nature and Uses*

Roy Underhill, Consultant
Television host of "The Woodwright's Shop", author of *The Woodwright's Shop, The Woodwright's Companion*, and *The Woodwright's Workbook*, and Master Housewright at Colonial Williamsburg

A Publication of
THE NATIONAL HISTORICAL SOCIETY

The Modern Carpenter Joiner and Cabinet-Maker presented up-to-date techniques and tools for its time. However, much has changed since 1902. Not all materials and methods described in these pages are suitable for the construction materials and tools of today. Before undertaking any of the building, remodeling or other practices described in these pages, the reader should consult with a reputable professional contractor or builder, especially in cases where structural materials may come under stress and where structural failure could result in personal injury or property damage. The National Historical Society, Cowles Magazines, Inc., and Cowles Media Company accept no liability or responsibility for any injury or loss that might result from the use of methods or materials as described herein, or from the reader's failure to obtain expert professional advice.

Library of Congress Cataloging-in-Publication Data

Survey of carpentrymasterpieces / G. Lister Sutcliffe, editor : Roy Underhill, consultant.

p. cm. — (The Modern carpenter joiner and cabinet-maker: v. 5)

Reprint. Originally published: 1902.

ISBN 0-918678-59-5

1. Roofs. 2. Floors. 3. Ceilings. I. Sutcliffe, G. Lister. II. Series.

TH2401.S87 1990 90-6198

694—dc20 CIP

CONTENTS

DIVISIONAL-VOL. V

SECTION VII.—CARPENTRY (*Continued*)

BY THE EDITOR

ILLUSTRATIONS

DIVISIONAL-VOL. V

ILLUSTRATIONS IN TEXT

SECTION VII.—CARPENTRY (*Continued*)

PREFACE

We are literally at the pinnacle of the carpenter's trade. A timber roof is strength and beauty combined, recognized in the 1734 *Builder's Dictionary* as "the most difficult and most useful Part of Carpenter's Work." The roof designs in this volume range from medieval to modern, from utilitarian to majestically decorative, and in most towns of any age at all, you will discover just such a diversity. Even within a single old building you can gauge the passage of the years by the technical changes in the framing of the roof.

Bruton Parish Church in Williamsburg, Virginia was built in 1720, with the steeple added in 1769. Climbing the stairs to the gallery, you pass through a little door into an upper world of magnificent and massive roof framing. Huge tulip poplar beams span the thirty-six foot breadth of the church. As simple as Stonehenge, they support the ceiling framing solely through the stiffness of their enormous cross-section. Such simplicity is possible only when nature can provide trees of unsurpassed strength and quality.

The principle rafters of the church are equally massive and simple, joined by high horizontal tie beams. Such tie beams exploit the tensile strength of wood to resist the outward thrust of the rafter feet. But not every complex roof needs to exploit tension. Even the magnificent hammer-beam roof is fundamentally a collection of shortened posts in compression. A hammer-beam roof could have all the pins pulled and tenons shorn, yet stand strong as ever. But fool with a tension truss and watch out! Walter Rose, in his memoir of the country carpenter's life, tells the story of a farmer's two overweight sons who tired of having to step over a tie beam crossing the loft of their barn. They decided to saw out a section of the offending beam to make a passageway. With the final stroke of the saw, the timber parted and the roof of their barn spread flat upon them, pushing the walls apart as it fell.

Walter Rose also recalled his father teaching him to lay out the cuts for roof framing by pacing them off with an iron square. He well remembered his father's pride in pointing out complex hipped roofs where "all the timber was sawn at the yard, each separate piece to the correct length and splay, without temporarily setting it up in position." Such prefabrication of timber frames at the carpenters' yard or "framing ground" meant that all the timbers had to be carefully numbered so that they could be reassembled at the site.

As with all such buildings, the timbers of the roof and steeple of Bruton Parish Church are chiseled and inscribed with Roman numerals. Erected in 1769 by Benjamin Powell, the steeple construction draws one relentlessly up ever steeper stairways, up ladders to the clock level, up through the network of timbers to where the bell hangs, and finally to the very top of the octagonal spire. As you climb, the timbers darken where two centuries of pigeon guano mingles with dust blown in on the cold wind that seeps in through the shingles. I have searched the steeple timbers in hope of finding the builder's signature. Inevitably though, the carved names that you discover are not of the builders, but of the sons of countless rectors of the church that have explored up here while their fathers tended business down below.

In the eighteenth and nineteenth centuries the traditional apprenticeship and father to son learning, began to be supplemented by carpenter's pattern books. These books spread new designs quickly through the world. Virtually every colonial gentleman's library boasted books of architecture and building. Working carpenters, too, began to depend on such books as *The Builder's Dictionary*, *The London Art of Building*, and *The Carpenter's Complete Instructor in Several Hundred Designs, consisting of Domes, Trussed Roofs, and various Coupolas*. The new designs developed hand-in-hand with an increasing scarcity of sizable timber and skilled labor. Framing elements became lighter and carpenters used more straps and iron bolts in the place of mortice and tenon joints.

Change though it might, the principles and challenge of timber roof framing will always be with us. To shelter, and to inspire. Even in his little cabin by Walden pond, Henry David Thoreau wrote of his dream of "a vast, rude, substantial, primitive hall, without ceiling or plastering, with bare rafters and purlins supporting a sort of lower heaven over one's head, . . . where the king and queen posts stand out to receive your homage."

ROY UNDERHILL
MASTER HOUSEWRIGHT
COLONIAL WILLIAMSBURG

SECTION VII—CARPENTRY (CONTINUED)

CHAPTER V

DESCRIPTION AND DELINEATION OF ROOFS

1. BUILDINGS WITH STRAIGHT SIDES

The most simple form of building is one erected on a rectangular plan. Such a building is roofed, generally, either with a roof of a single slope, called a *shed* or *lean-to roof*, as in No. 1, fig. 655, the wall of one of the sides of the building being carried so much higher than the wall parallel with it as to give the required slope to the roof,—or with a roof of double slope, as in No. 2, known as a *span roof*. In the latter, the planes forming the slopes are equally inclined to the horizon; the meeting of their highest sides makes an arris, which is called *the ridge* of the roof; and the triangular spaces in the end walls are called *gables*.

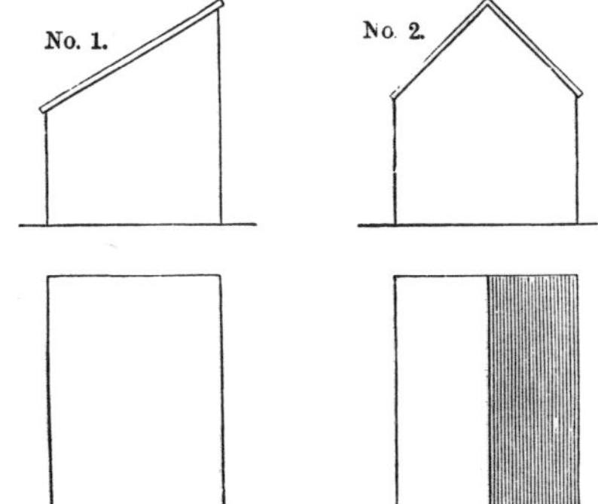

Fig. 655.—Plans and Elevations of Roofs No. 1, Shed or lean-to; No. 2, Span roof.

Sometimes the slopes of a span roof are unequal. This form is often used for weaving sheds (the steeper slope being glazed), and may be known as a *glazed shed roof*. The slopes are generally 30° and 60° respectively, thus forming a right angle at the ridge.

When a building is erected on a rectangular plan, of which the four sides are equal, it may be covered with a roof of one or two slopes, as above. But it may happen that no necessity exists for making any of the opposite pairs of sides gables; or there may be reasons why all the sides should be gables. In the latter case, two roofs of equal slopes intersect each other. This roof (fig. 656) has two ridges *a b*, *c d*, and four hollow arrises *fe*, *ge*, *he*, *ke*, made by the intersections of the planes of the slopes, and lying over the diagonals of the square. The arrises are termed *valleys* or *flanks*. In the former case, a mode much more simple, and often preferable, because simpler in construction, is to make the sides of the roof spring from the sides of the square with an equal slope. The result (fig. 657) is a pyramid, and the intersections of the sloping planes form salient angles, or arrises, which are known as *hips*. This kind of roof is called a pavilion roof. If its base is a polygon, of which the sides are equal, the pyramid will be composed of as many equal triangles as the polygon has sides.

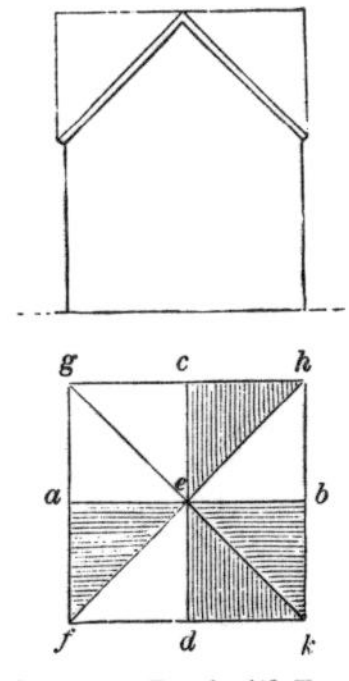

Fig. 656.—Roof with Four Gables and Four Valleys

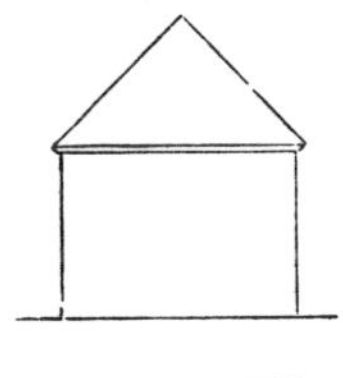

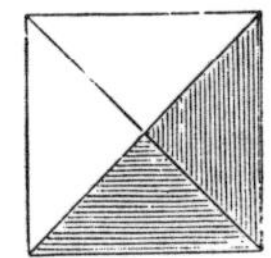

Fig. 657.—Pyramidal or Pavilion Roof with Four Hips

When the sides of a parallelogram on which a building is raised are not very unequal, it may be roofed with a pavilion roof, as in fig. 658; the slopes on corresponding and opposite sides being equal, but those on contiguous sides different.

In the roof (fig. 659) the sides *a b*, *c d*, are truncated by sides of the same slope rising from the ends *c a*, *d b*. These form, by their meeting with the former, the arrises or

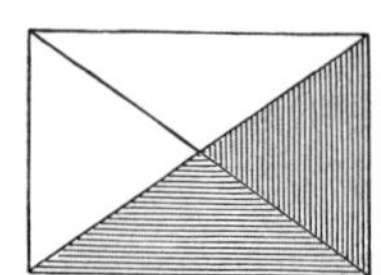

Fig. 658.—Plan of Oblong Pavilion Roof

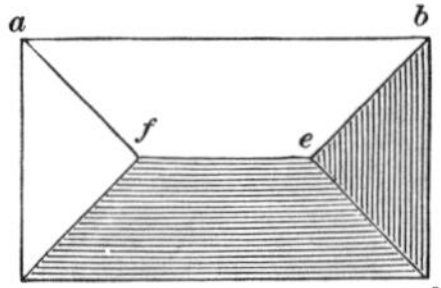

Fig. 659.—Hipped Roof

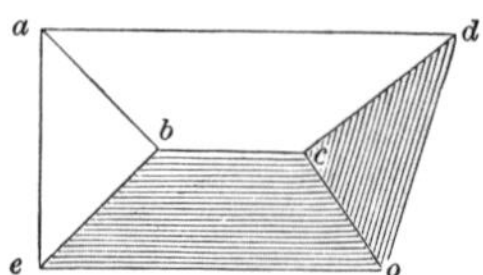

Fig. 660.—Hipped Roof with One End Skewed

hips *af*, *cf*, *b e*, *d e*. The roof is called a hipped roof, and the rafters on the lines of the arrises are called hip-rafters. When the end of such a roof is at right angles to its side, as *abe* (fig. 660), it is called a right hip; when the angles are unequal, as at *dco*, it is an oblique or skewed hip.

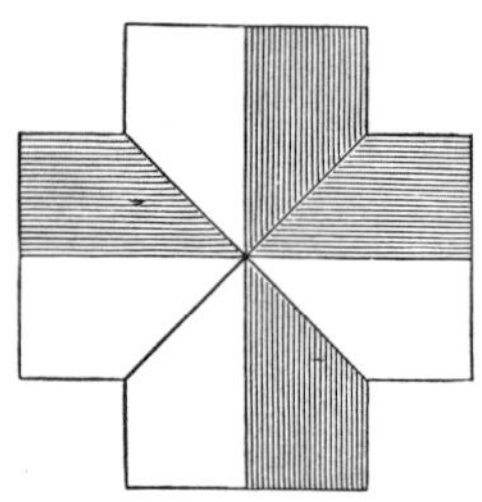

Fig. 661.—Cross-shaped Roof with Four Gables

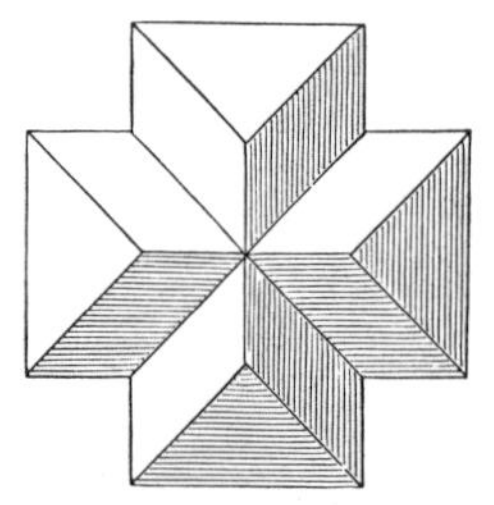

Fig. 662.—Cross-shaped Roof with Four Hips

When the plan of a building is composed of two equal parallelograms crossing each other at right angles, each of the parallelograms may be covered with a roof of two slopes and two gables, as in fig. 661, or with a hipped roof, as in fig. 662; in each case forming four valleys at their intersections.

When the intersecting parallelograms are unequal in length, the shortest may be roofed pyramidally, the slopes of its contiguous sides being unequal, and the longest with a ridge, as

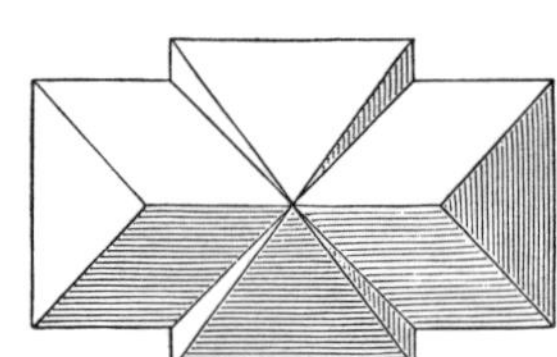

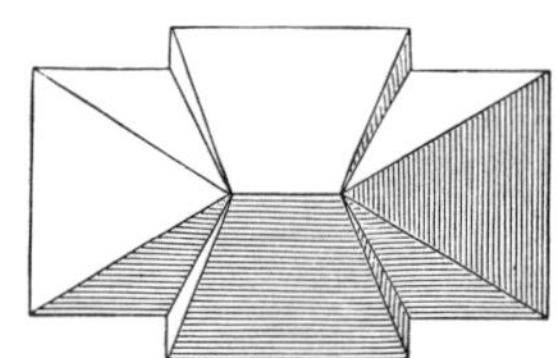

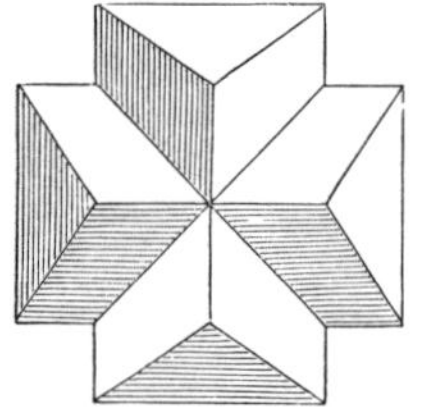

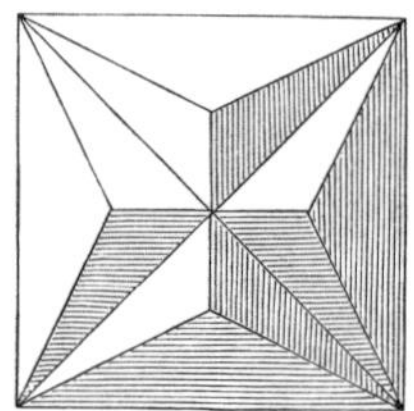

Figs. 663–666.—Intersecting Hip Roofs with Unequal Slopes

in fig. 663; or there may be a short ridge common to both, as in fig. 664. In both cases the roof-slopes are unequal. Figs. 665 and 666 are plans of roofs having valleys, hips, and ridges, the slopes in both being unequal.

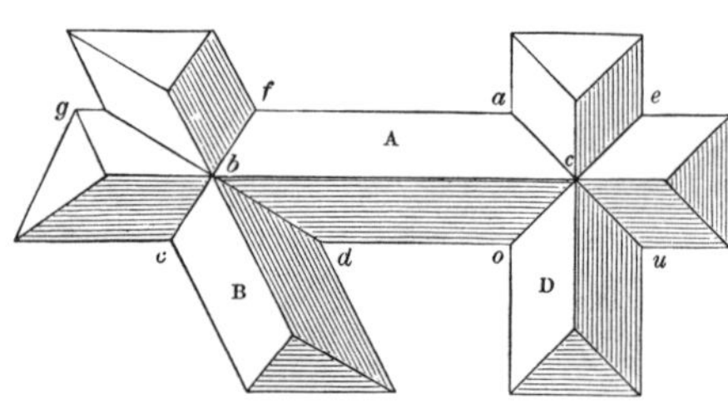

Fig. 667.—Oblique and Square Intersecting Roofs

In fig. 667 is represented the horizontal projection or plan of the roofs of a building composed of three ranges crossing each other. Their six extremities, square or skewed, are hipped, according to the form of the plan. The valleys *c a*, *c e*, *c o*, *c u* are equal, because the buildings A and D are equal in width, and cross each other at right angles. The valleys *b c*, *b d*, *bf*, *b g* are unequal, because the angles formed by the intersecting buildings are not 90°. In this combination of roofs the hips occur together in pairs, and the valleys four and four.

In the next figure (fig. 668) the ranges of building enclose a court. The external slopes form hips at the angles; the internal slopes meet at valleys. All the hips divide the angles to which they correspond into two equal parts. In the same way the valleys divide the interior angles into equal parts; and in general the hips and valleys are equal when they result from the meeting of two ranges of buildings of the same width, with roofs of equal height; but they become irregular when the buildings are of unequal width, though of the same height, or when the opposite sides of the roofs are of different slopes. At the range of building M in the figure, which meets the range A at right angles, the slopes, and consequently the valleys, are unequal, the inequality being proportionate to the deviation from the dotted lines; and the gable is also irregular, as shown

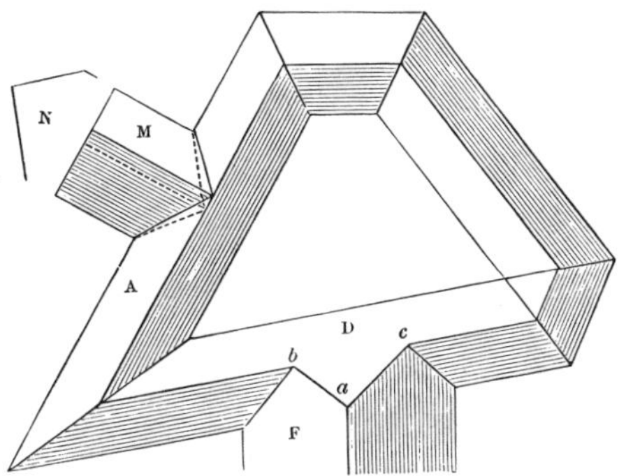

Fig. 668.—Roof of Buildings around irregular Court

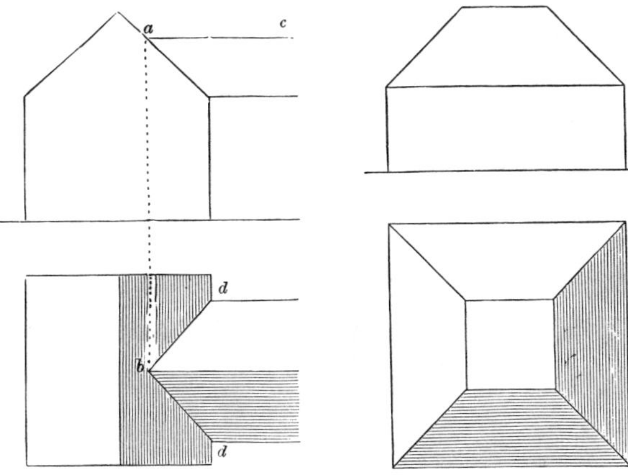

Fig. 669.—Intersection of Roofs of Equal Slope and Unequal Span

Fig. 670.—Pavilion Roof with Platform

by the section N. When the greater width of a building causes its roof to rise higher than the roof which it meets, as the roof of the wide range F meeting D, the connection is completed by extending the slope of D so as to truncate the summit of F, and form the hips *b a*, *c a*.

When two buildings of unequal width meet each other, and the ridges of the roofs are not kept of equal height, as in fig. 669, the horizontal projection or plan of the lower roof is found by drawing a line *a b* from the point in the vertical projection *a*, where the line of the ridge *a c* of the lower roof meets the slope of the higher roof, to the seat of the ridge on the plan below, and joining *b d*, *b d* for the valleys formed by the intersections. When the slopes are equal, as in the figure, *d b d* will be a right angle.

In fig. 670 is shown a pavilion roof truncated by a plane parallel to its base; it consists, therefore, of four slopes and a platform.

In fig. 671, A is the horizontal, and B the vertical projection of a square building, presenting four equal gables and as many equal slopes; but the intersections of the slopes, in

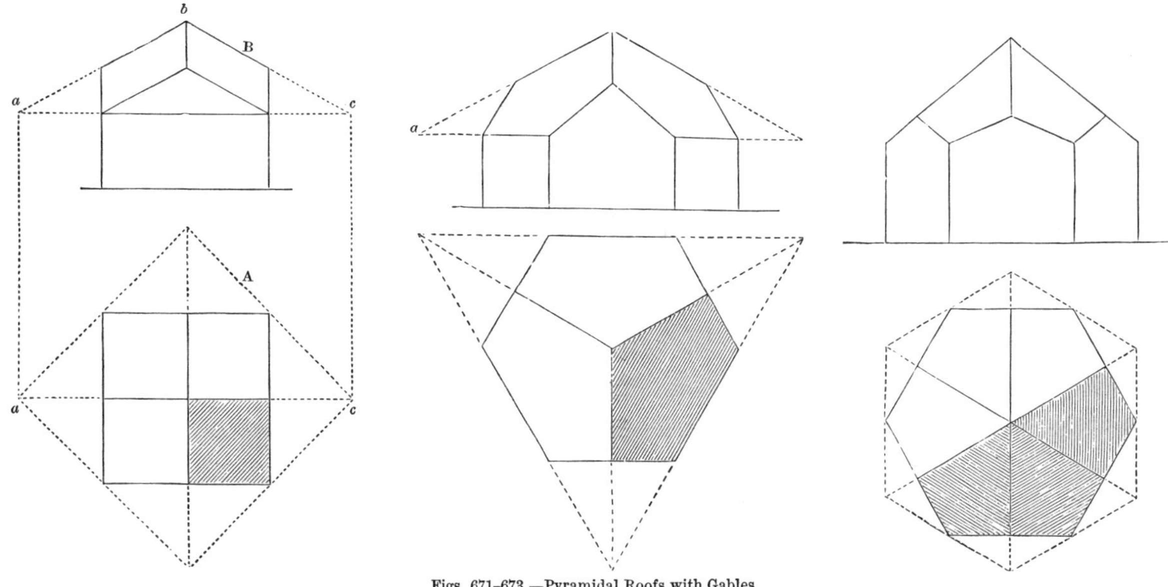

Figs. 671–673.—Pyramidal Roofs with Gables

place of lying over the diagonals of the square, connect together the summits of the gables. This roof, it will be seen by the dotted lines, is the section of a pyramidal roof *a b c*, made by four planes parallel to its diagonal. The towers of Romanesque buildings are often roofed in this manner, the pitch, however, being very much quicker than that shown in the illustration.

The combination of hips and gables may be used for figures of any number of sides. If the number of sides is even, the sides may be alternately gabled and horizontal, as in fig. 672, the plan of which is a hexagon, the result of the truncation of a triangular pyramid by planes parallel to its opposite sides. Where each side has a gable, as in fig. 673, the resulting figure is a hexagonal pyramid truncated by six planes forming gabled sides.

In fig. 674, A is the horizontal projection of a roof with hips or pavilion ends, truncated by vertical planes; B is the vertical projection of the side, and C that of the end. It is known as the *half-hip*, or *hip and gable*.

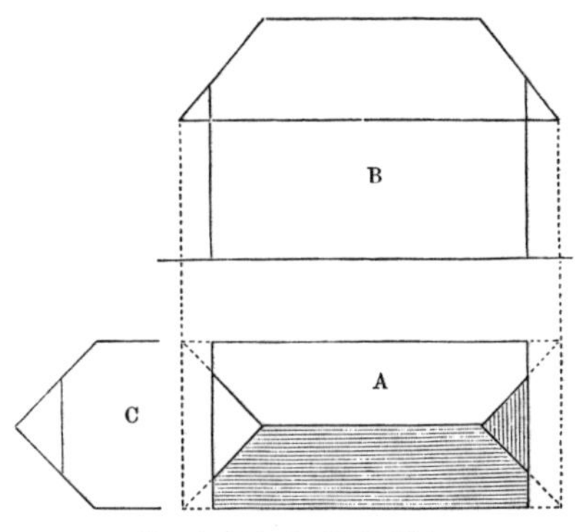

Fig. 674.—Roof with Half-hips

When it is desirable to keep the roofing over a wide building of a rectangular plan low, it may be effected by dividing the span into two, with four principal slopes, two external and two internal. This, which produces a section somewhat resembling the letter **M** (fig. 675), is, from its form, called *an* **M**-*roof*. At the meeting of the interior slopes is formed a gutter for the water. Fig. 676 is the plan of such a roof with gables, and fig. 677 the same with hips. In order to avoid the long gutter, another roof may be advantageously introduced, as in fig. 678, crossing between the two ridges at right angles, and forming valleys by the intersection of its slopes with the interior slopes of the longitudinal roofs. For the sake of external appearance, or to collect the water in the centre, two cross roofs are sometimes used, enclosing a central space as in fig. 679; this is called *a hopper roof*.

Sometimes the plan of a building is irregular, and its sides are not parallel. If the roof is constructed so that the ridge slopes downwards to the narrow end, its sides will be planes;

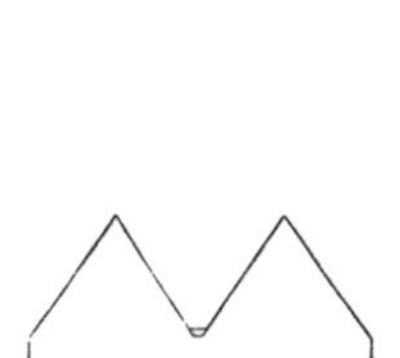
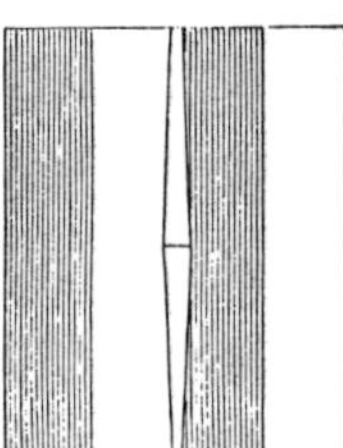
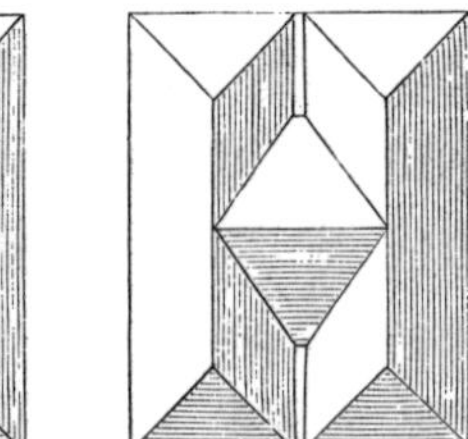
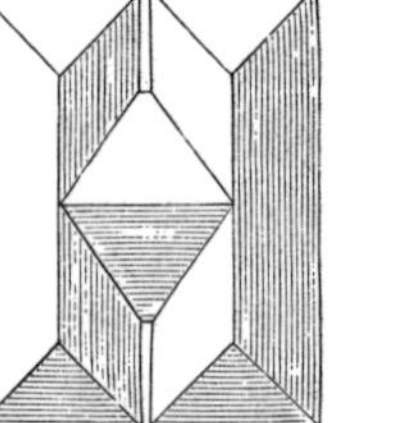

Figs. 675–678.—Section and Plans of **M**-roofs

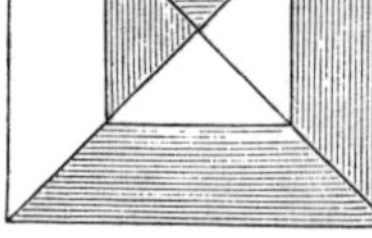

Fig. 679.—Hopper Roof

this will also be the case if the ridge is horizontal, and the eaves are made to slope downwards to the broad end; but if the ridge and eaves are made horizontal throughout, the sides of the roof will become twisted or winding.

Let $a\,d\,d'\,a'$ (fig. 680) be the plan of an irregular building. The roof is hipped, having sloped ends forming the two triangles $a\,p\,a'$, $d\,g\,d'$, which may be isosceles; and the ridge is projected on the line pg, which is not parallel to either of the sides $a\,d$, $a'd'$. Two cases here present themselves: the ridge projected on $p\,g$ is either horizontal, and its extremities are determined by its meeting with the planes of the sloping ends—the larger sides in this case being twisted; or it has an inclination determined by the intersection of the planes $a\,p\,g\,d$, $a'p\,g\,d'$, springing with equal slopes from the wall head. It is easy to observe that the ridge will not in either case appear parallel to the faces of the building. The two greatest sides of the roof are surfaces generated by a line which, moving from $a\,d$ or $a'd'$ to meet the ridge $p\,g$, is kept in contact with a vertical line passing through c or c', where the arrises of the hips meet when produced.

In fig. 681, two equally inclined planes spring from $a\,d$, $a'd'$, and their intersection produces an inclined ridge *pf*. The points *qq* are taken at the level of *p*, and the arris *fq* is thus symmetrical with *fp*. The triangle *fqq* is a platform, the slope of which will vary in proportion to the pitch of the roof.

The same plan may be covered by a roof with horizontal curbs, as in fig. 682. The four sides of the roof are planes, springing at the same inclination from the walls *abcd*. The

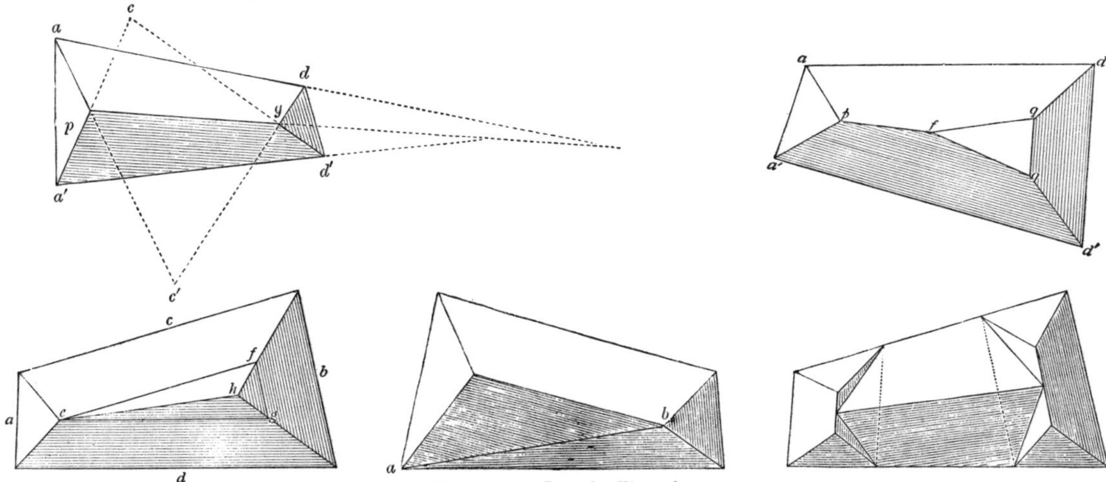

Figs. 680-684.—Irregular Hip-roofs

curbs *ef*, *fg*, *ge* are continued at the height determined by the intersection of the hips at *e*; and a pyramidal construction *efg* is added at so low a pitch as not to be visible from the ground. In place of these pyramidal roofs, the three sides of the building *cbd* may have roofs of two slopes intersecting at the lines *eh*, *hf*, *hg*, forming what is called an irregular hopper roof; or the part *efg* may be formed into a lead flat, and this will generally furnish the best way out of the difficulty.

In fig. 683 another method of preserving the ridge horizontal, without twisting the sides, is shown. In this case one of the sides is roofed in two slopes, forming an arris *ab*, the part above this arris having a smaller slope than that below. This arrangement cannot be recommended.

In fig. 684 another method of roofing the same space is shown.

Fig. 685 shows various methods of roofing a trapezium. In A the sides are planes, and form an irregular pavilion roof with unequal slopes. In B planes of the same slope rise to

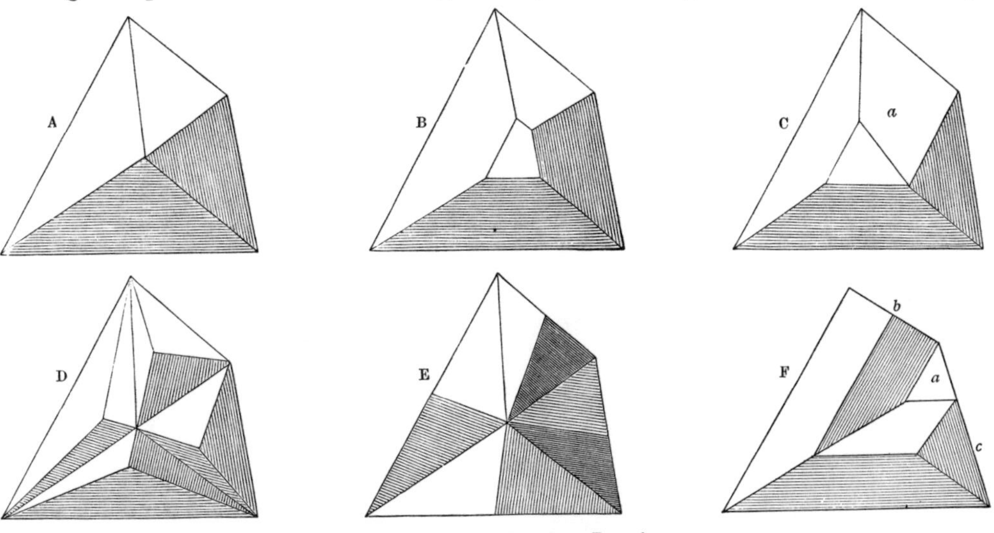

Fig. 685.—Six Methods of Roofing a Trapezium

the same height, and are united by a platform. In C the slopes are not equal, and one side *a* is twisted. In D two irregular pavilion roofs cross each other, forming valleys at their intersections. In E there are four gables and four valleys. If the ridges in D and E are level the slopes must be very unequal, and the effect will be unsatisfactory. B is the best arrangement shown; but other arrangements are possible, such as the hopper roof or that given in F, where two equal-sided roofs meet at an acute angle, and a small lead flat occupies the remaining space *a*; the ends may be gabled as at *b*, or hipped as at *c*.

2. BUILDINGS WITH CURVED SIDES

Fig. 686 shows a roof over an oval building; the ridge is straight and horizontal, and consequently the sides are twisted; but in this case the appearance of the sides is not

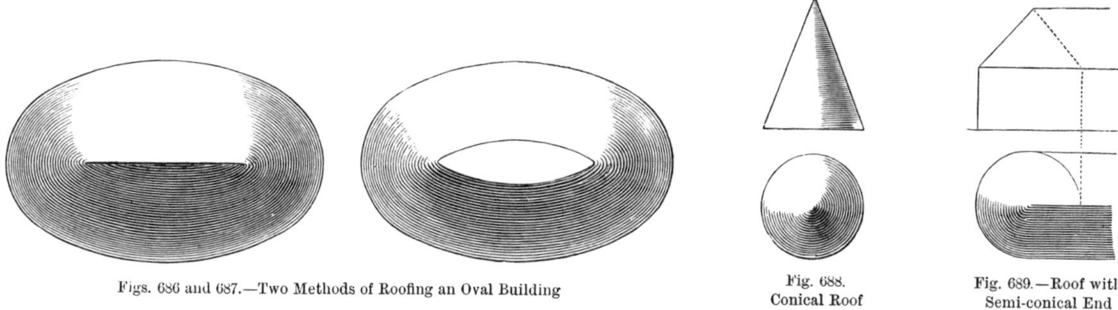

Figs. 686 and 687.—Two Methods of Roofing an Oval Building

Fig. 688. Conical Roof

Fig. 689.—Roof with Semi-conical End

disagreeable. In fig. 687 the sides are not twisted, but slope everywhere alike, and the roof is truncated and terminated by a platform. This is the better arrangement.

Fig. 688 is a conical roof.

Fig. 689 is a roof over a rectangular plan, with a semicircular end. The end of the roof is consequently a semi-cone.

Fig. 690 is an annular roof.

Fig. 691 is the roof of a crescent building, with one end gabled and the other hipped.

Fig. 692 is the vertical and fig. 693 the horizontal projection of two united conical roofs of equal

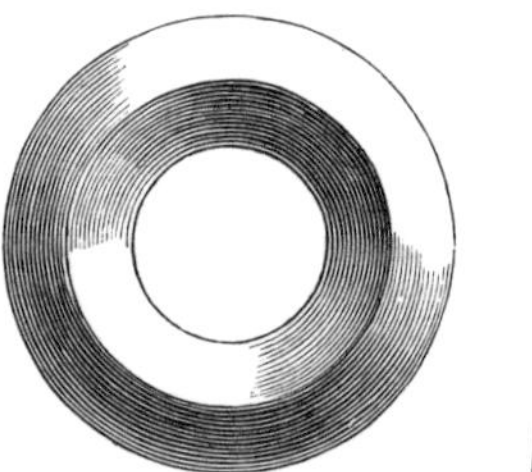

Fig. 690.—Annular Roof

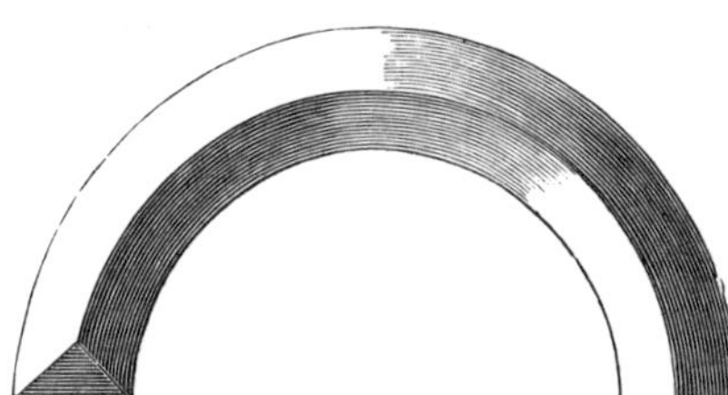

Fig. 691.—Crescent Roof

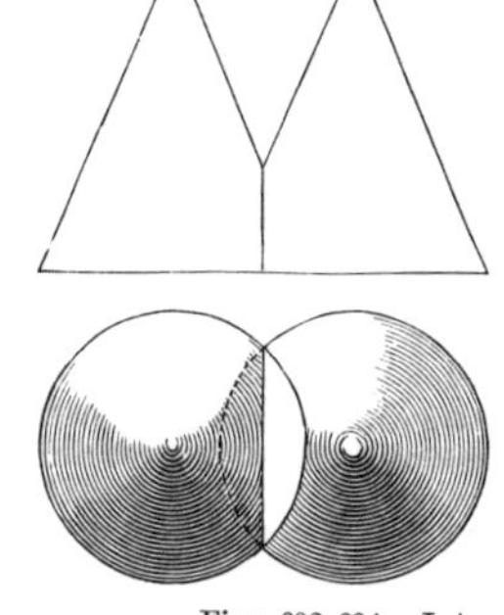

Figs. 692–694.—Intersecting Conical Roofs

size. The line of intersection is a hyperbolic curve, which appears as a straight line in the two projections given.

Fig. 694 shows the junction of a large and a small conical roof. The line of intersection will vary according to the relative slopes of the two cones.

Fig. 695 is the junction of a conical and a pavilion roof.

Fig. 696 is the junction of a span roof with a large conical roof, and fig. 697 that of a conical and a rescent roof.

Figs. 695 and 696.—Intersecting Conical and Plane Roofs

Fig. 697.—Intersecting Conical and Crescent Roofs

3. SPIRES

In fig. 698, A and B are the horizontal and vertical projections of a round pavilion, the roof of which is formed by the truncating of a cone by four inclined planes; the curves of the arrises are portions of ellipses.

In fig. 699, A is a horizontal projection of a roof, formed by the setting of a square pyramidal roof diagonally on a pavilion roof of lower elevation. B is the vertical projection

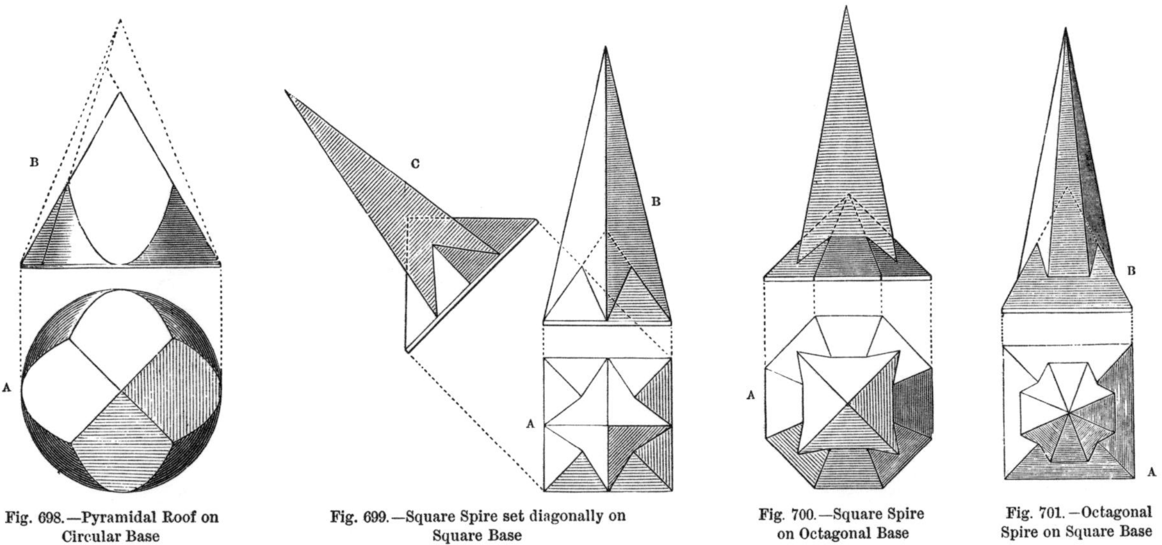

Fig. 698.—Pyramidal Roof on Circular Base

Fig. 699.—Square Spire set diagonally on Square Base

Fig. 700.—Square Spire on Octagonal Base

Fig. 701.—Octagonal Spire on Square Base

on a plane parallel to one of the faces of the lower pyramid, and C the vertical projection on a plane parallel to its diagonal.

In fig. 700, A is the horizontal, and B the vertical projection of a pyramidal roof with a square base set on an octagonal pavilion roof of lower elevation.

In fig. 701, A is the horizontal, and B the vertical projection of a pyramidal roof with an octagonal base, set on a pavilion roof of lower pitch with a square base. This is a common

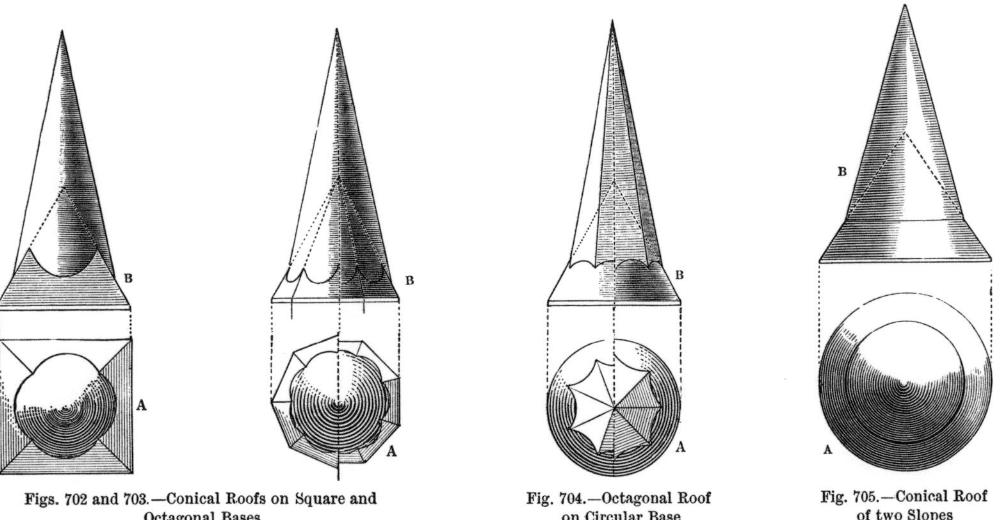

Figs. 702 and 703.—Conical Roofs on Square and Octagonal Bases

Fig. 704.—Octagonal Roof on Circular Base

Fig. 705.—Conical Roof of two Slopes

form of roof for Gothic spires, both of timber and stone, and is known as the "broach" spire. The four cardinal sides of the octagon usually rise directly from the four walls of the tower.

In fig. 702, A and B are the horizontal and vertical projections of an acute conical roof set on a square pyramid.

In fig. 703, A and B are the horizontal and vertical projections of an acute conical roof set on an octagonal pyramid.

In fig. 704, A is the horizontal, and B the vertical projection of a roof formed by an octagonal pyramid set on a cone. This is a very common form of roof in Belgian and French *chateaux*, but the change from the octagon to the cone is generally "humoured" to an easy curve.

In fig. 705, A is the horizontal, and B the vertical projection of a roof formed by an acute cone set on one of lower pitch.

4. CURB OR MANSARD ROOFS

To diminish the excessive height of roofs, their sharp summit is sometimes suppressed and replaced by a roof of a lower slope. These roofs have the advantage of giving ample attic space with a smaller height than would be required by a V-roof. They are variously known as "curb" or "gambrel" roofs, and "Mansard" roofs, the latter name being usually confined to those roofs in which the lower slopes form angles of not less than 60° with the horizontal plane, while roofs of smaller pitch are known as "curb" or "gambrel" roofs.

François Mansart, who died in 1666, brought this sort of roof into fashion in France, and was for a long time regarded as its inventor. Bullet, however, says that Mansart truncated his roofs after the example of one at Chilly, by Metezau. Mesanges asserts that he took the idea from a frame composed by Sangallo, and that Michael Angelo employed it in the construction of the dome of St. Peter's; but Krafft, in his work on *Carpentry*, states that houses in Lower Brittany were covered with these roofs in the latter part of the fifteenth century. The roof is usually known, however, as the Mansard roof, and the garrets formed in such roofs were called *Mansards*.

The Mansard roof may be described in several ways:—

1st. Fig. 706.—The triangle *a d b* represents the profile of a high-pitched roof, the height being equal to the base, and the basal angles being therefore 60° each. At the point *e*, in the middle of the height *c d*, draw a line horizontally *h e i*,

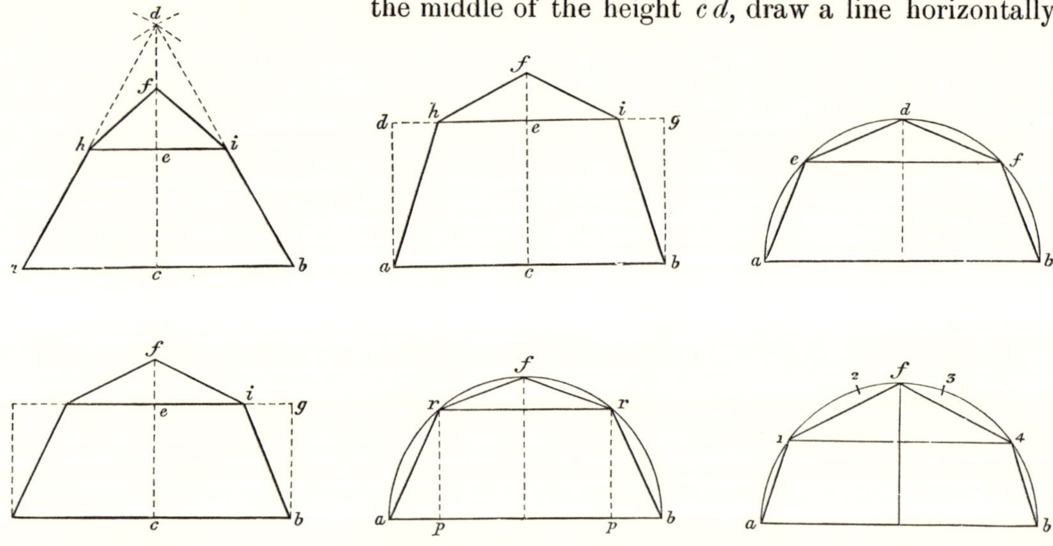

Figs. 706–711.—Methods of describing Mansard Roofs

parallel to the base *a b*, to represent the upper side of the tie-beam, and make *e f* equal to the half of *e d*; then *a h f i b* will be the profile of the Mansard roof.

2nd. Fig. 707.—Make *c e*, the height of the lower roof, equal to half the width *a b*, and construct the two squares *a d e c*, *c e g b*; also make *d h*, *e f*, and *g i* each equal to one-third of the side of either square; then will *a h f i b* be the profile required.

3rd. Fig. 708.—On the base *a b* draw the semicircle *a d b*, and divide it into four equal parts, *a e*, *e d*, *d f*, *f b*; join the points of division, and the resulting semi-octagon is the profile required. The slopes of the upper roof form angles of only $22\frac{1}{2}°$, and this roof is therefore considerably less than "quarter-pitch", and would be unsuitable for covering with slates, tiles, shingles, &c.

4th. Fig. 709.—Whatever be the height of the Mansard *c e* or *b g*, D'Aviler makes *g i* equal to the half of that height, and the height *e f* of the false roof equal to the half of *e i*. The upper roof, therefore, is exactly "quarter-pitch".

5th. Fig. 710.—Describe the semicircle *a f b*, and divide each half of the base *a b* into three equal parts. From the last divisions *p p* the perpendiculars *p r*, *p r* are erected, cutting the semicircle in *r r*; then *a r f r b* is the profile. This also gives a low pitch to the upper roof.

6th. Fig. 711.—Describe a semicircle on the base *a b*, and divide its circumference into five equal parts, in 1 2 3 4; then the chords *a* 1, *b* 4 are the sides of the lower roof, and 1 *f*, 4 *f* those of the upper roof. This is a convenient method, and makes the upper roof almost exactly "quarter-pitch".

The form of the Mansard roof, it will be seen, may be infinitely varied, according to the fancy of the designer, the purposes for which the roof-space is required, and the nature of the roof-covering. In many cases the lower slopes are made of curved outline; examples of these will be given in the next chapter.

5. HIPS AND VALLEYS

In its most simple form the *hip-roof* is a quadrilateral pyramid, each angle being a *hip*, and the rafter in each angle a *hip-rafter*. The *common rafters* which lie between the hip-rafters in the planes of the sides of the roof, and which, by abutting on the hip-rafters, are necessarily shorter than the length of the sloping side, are called *jack-rafters*.

The points to be determined in a hip-roof are these, viz.:—

1. The angle which a common rafter makes with the plane of the wall-head; that is, the angle of the slope of the roof.
2. The angle which the hip-rafters make with the plane of the wall-head.
3. The angles which the hip-rafters make with the adjoining planes of the roofs. This is called the backing of the hip.
4. The height of the roof.
5. The lengths of the common rafters.
6. The lengths of the hip-rafters.
7. The length of the wall-plate contained between the hip-rafter and next adjacent entire common rafter.

The first, fourth, fifth, and seventh of these are generally given, and all the others can be found from them by construction.

The plan of a building and the pitch of the roof being given, to find the lengths of the rafters, the backing of the hips, and the shoulders of the jack-rafters and purlins.

Let A B C D (fig. 712) be the plan of the roof, A D being parallel to B C.

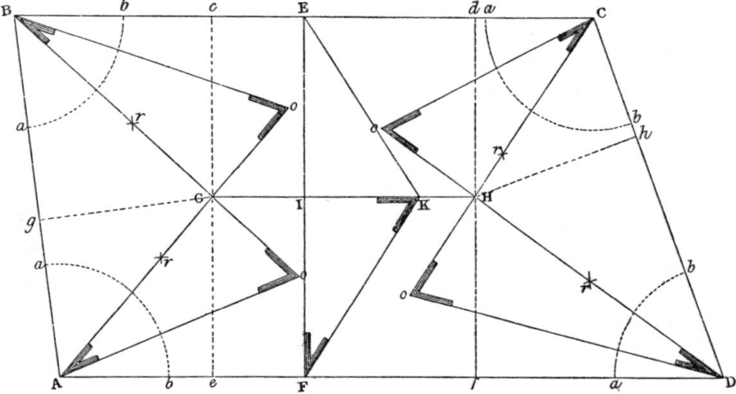

Fig. 712.—Method of finding Lengths of Hip-rafters in Irregular Hip-roof

Draw G H parallel to the sides A D, B C, and in the middle of the distance between them. From the points A, B, C, D, with any radius, describe the curves *a b*, *a b*, cutting the sides of the plan in *a* and *b*. From these points describe equal arcs to bisect the four angles of the plan, and from A B C D, through the points *r r r r* thus formed, draw the lines of the hip-rafters A G, B G, C H, D H, cutting the ridge line G H in G and H, and produce them indefinitely.

The dotted lines *c e, d f* are the seats of the last entire common rafters on the sides of the building, and *g* G, *h* H are the only entire common rafters on the hipped ends.

Through any point I in the ridge line draw E I F at right angles to G H. Make I K equal to the height of the roof, and join E K, F K: then E K is the length of a common rafter.

From G draw G *o*, G *o* at right angles to B G and A G, and from H draw H *o*, H *o* at right angles to C H and D H. Make G *o*, H *o* equal to I K, the height of the roof; and join A *o*, B *o*, C *o*, D *o* for the lengths of the hip-rafters. If the triangles A *o* G and B *o* G be turned round their seats A G, B G, until their planes are perpendicular to the plane of the plan, the points *o o* and the lines G *o*, G *o* will coincide, and the rafters A *o*, B *o* be in their true positions, as shown in plan at A G, B G.

The method of finding the backing of a hip-rafter will be described hereafter (fig. 716).

Let A B C D (fig. 713) *be the plan of an irregular roof, in which it is required to keep the ridge level.*

Bisect the angles at the two ends by the lines A *l*, B *l*, C G, D G, in the same manner as before; and through G draw the lines G E, G F parallel to the sides C B, D A respectively, cutting B *l* and A *l* in E and F; join E F: then the triangle E G F is a flat, and the remaining triangles and trapeziums are the inclined sides.

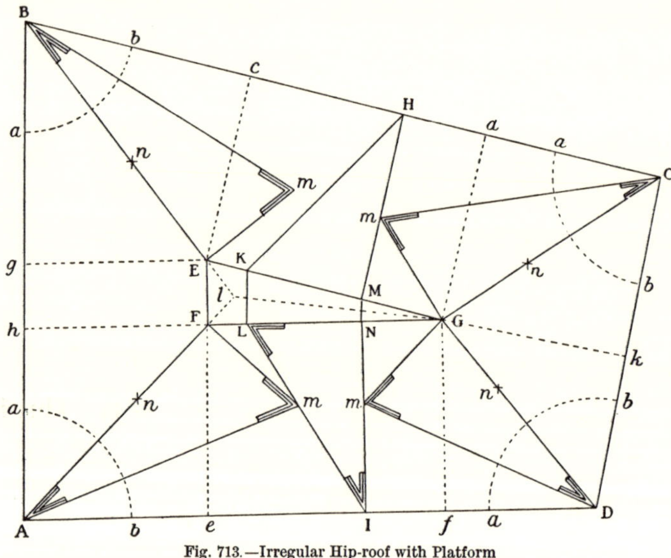

Fig. 713.—Irregular Hip-roof with Platform

Join G *l*, and draw M N perpendicular to it: from the points M and N draw M H and N I perpendicular to G E and G F respectively, and set off along M E and N F the distances M K and N L equal to the height of the roof: then draw H K, I L, and these will be the lengths of the common rafters.

At E, set up E *m* perpendicular to B E; make it equal to M K or N L, and join B *m* for the length of the hip-rafter; and proceed in the same manner to obtain A *m*, C *m*, D *m*.

From the summit of each hip-rafter, let fall perpendiculars to the two adjacent sides, namely, E *g*, E *c*, G *d*, G *k*, G *f*, F *e*, and F *h*; these represent the last entire common rafters.

To find the hip and valley rafters of a compound irregular roof (fig. 714).

In the compound roof shown by the plan, in which the ridge is level throughout although the buildings are of different widths, the method of proceeding to find the hip and valley rafters of the right-lined parts of the roof is the same as in the two former cases, and will be evident on inspection.

In the circular part, proceed as follows:—Draw *c d*, a radius to the curve, as the seat of one pair of the common rafters *c b*, *d b*, and bisect it in *a*: through *a* describe the concentric curve *k a* W *n a*, which is the seat of the circular ridge: produce the lines of the other ridges to meet this curved line in *a* W *k*, and connect the angles of the meeting roofs with these points, as in the drawing: divide the seat of one pair of the common rafters in each roof, as X Y, P Q, T U, and *e f*, into the same number of equal parts; and through the points of division draw lines parallel to the sides of their respective roofs, intersecting the curved lines drawn through the points of the curved roof; and through the points of intersection draw the curves Z *l m a*, D *l m* W, &c., which give the lines of the hips and valleys.

On C *a*, the meeting of the left-hand roof with the circular roof, erect the perpendicular *a b* at *a*, and make it equal to the height of roof; and join C *b*, for the length of the valley-rafter: proceed in the same manner for the hip-rafter Z *b*, and for the other hip and valley rafters. This method assumes that the rafters will form straight lines on plan, and not

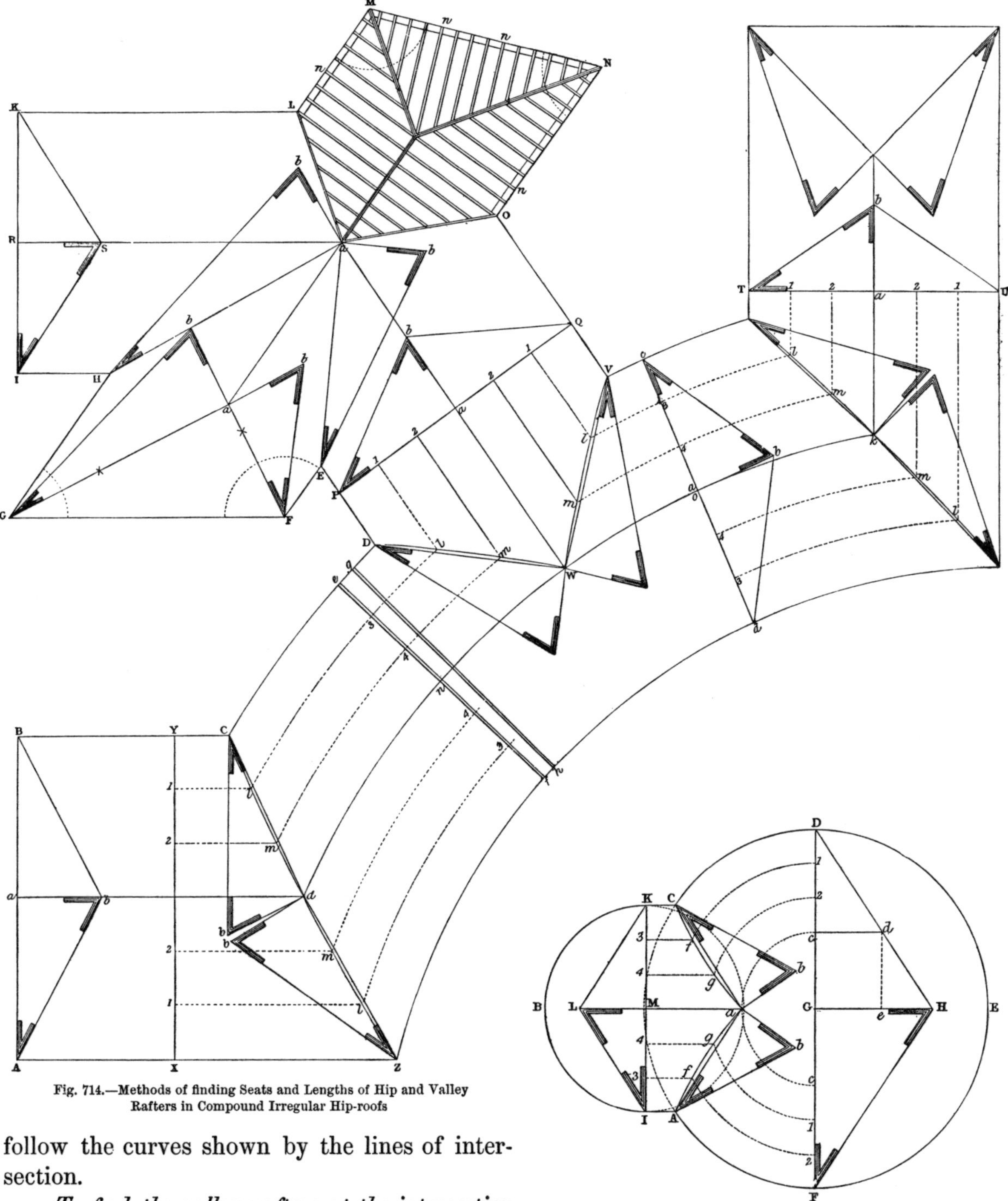

Fig. 714.—Methods of finding Seats and Lengths of Hip and Valley Rafters in Compound Irregular Hip-roofs

Fig. 715.—Valley-rafters at Intersection of Span and Conical Roofs

follow the curves shown by the lines of intersection.

To find the valley-rafters at the intersection of the roof B *with the conical roof* E (fig. 715).

Let D H, F H be the common rafters of the conical roof, and K L, I L the common rafters of the smaller roof, all of the same pitch.

On G H set up G *e* equal to M L, the height of the span roof, and draw *e d* parallel to D F, and from *d* draw *c d* perpendicular to D F. The triangle D *d c* will then by construction be equal to the triangle K L M, and will give the seat and the length and pitch of the common rafter of the smaller roof B. Divide the lines of the seats in both figures, D *c*, K M, into the same number of equal parts; and through the points of division in E, from G as a centre, describe the curves *c a*, 2*g*, 1*f*, and through those in B, draw the lines 3*f*, 4*g*, M *a* parallel to the sides of the roof, and intersecting the curves in *f g a*. Through these points trace the curves C*f g a*, A*f g a*, which give the lines of intersection of the two roofs. Then to find the valley-rafters, join C *a*, A *a*; and on *a* erect the lines *a b*, *a b* perpendicular to C *a* and A *a*, and make them respectively equal to M L; then C *b* and A *b* are the lengths of the valley-rafters, assuming them to be fixed to form straight lines on plan, and not made to the curves shown by the lines of intersection.

To find the backing of a hip-rafter.

1.—*Where the angles of the building are right angles* (fig. 716).

Let B *b*, *b* C be the common rafters, A D the width of the roof, and A B equal to one-half the width. Bisect B C in *a*, and join A *a*, D *a*. From *a* set off *a c*, *a d*, equal to the height of the roof *a b*, and join A *d*, D *c*; then A *d*, D *c* are the hip-rafters. To find the backing: from any point *h* in A *d*, draw the perpendicular *h g*, cutting A *a* in *g*; and through *g* draw perpendicular to A *a* the line *e f*, cutting A B, A D in *e* and *f*. Make *g k* equal to *g h*, and join *k e*, *k f*; the angle *e k f* is the angle of the backing of the hip-rafter A *a*, as shown in section at O.

2.—*Where the sides of the building are parallel, but the angles are not right angles* (fig. 717).

Bisect A D in *a*, and from *a* describe the semicircle A *b* D; draw *a b* parallel to the sides A B, D C, and join A *b*, D *b*, for the seat of the hip-rafters. From *b* set off on *b* A, *b* D, the lengths *b d*, *b e*, equal to the height of the roof *b c*, and join A *e*, D *d*, for the lengths of the hip-rafters. To find the backing of the rafters:—In A *e*, take any point *k*, and draw *k h* perpendicular to A *e*. Through *h* draw *f h g* perpendicular to A *b*, meeting A B, A D in *f* and *g*. Make *h l* equal to *h k*, and join *f l*, *g l*; the angle *f l g* is the backing of the hip. Proceed in a similar manner for the rafter D *d*.

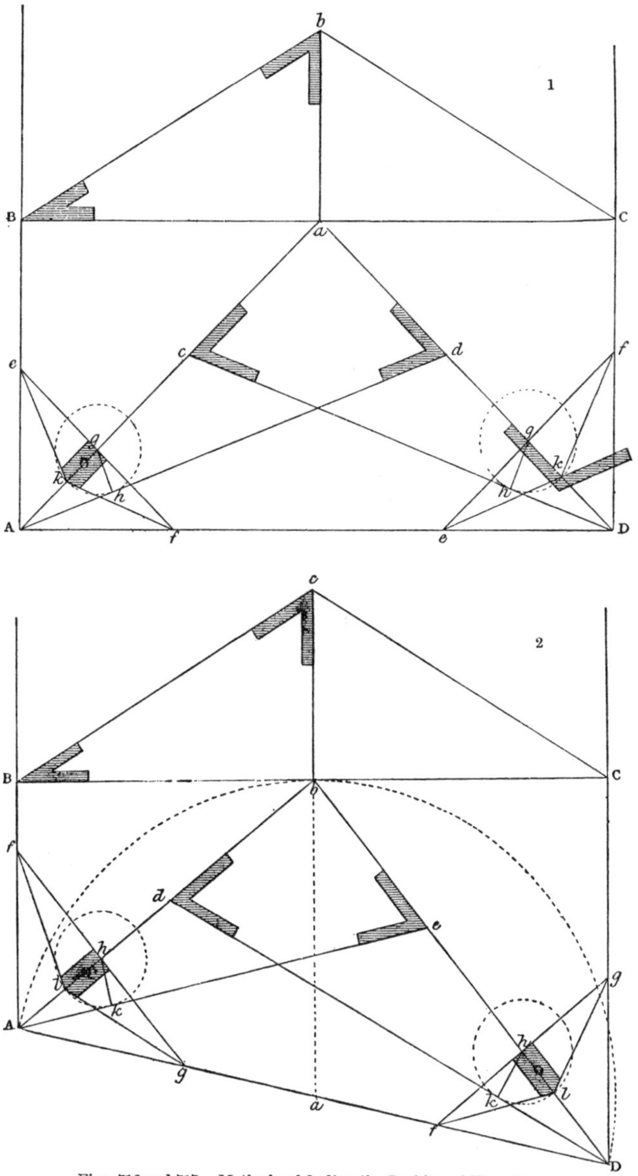

Figs. 716 and 717.—Methods of finding the Backing of Hip-rafters

To find the bevel of the shoulder of the purlins.

Fig. 718.—*Where the purlin has one of its faces in the plane of the roof*, as at E. From *c* as a centre, with any radius, describe the arc *d g*; and from the opposite extremities of the diameter, draw *d h*, *g m* perpendicular to B C. From *e* and *f*, where the upper adjacent sides of the purlin produced cut the curve, draw *e i*, *f l* parallel to *d h*, *g m*; also draw *c k* parallel to *d h*. From *l* and *i* draw *l m* and *i h* parallel to B C, and join *k h*, *k m*. Then *c k m* is the down bevel of the purlin, and *c k h* is its side bevel.

Where the purlin has two of its sides parallel to the horizon. This simple case is shown worked out at F (fig. 718). It requires no explanation.

Where the sides of the purlin make various angles with the horizon. Fig. 719 shows the application of the method described in fig. 718, to these cases.

To find the length and bevel of the jack-rafters.

Fig. 720, No. 1.—Let A B C D be the plan. Draw the hipped end of the roof A E D in any of the manners already described.

Bisect B C in E; and from E as a centre, with the radius E C, describe the semicircle C I *d*. Through E draw *d e* parallel to A D. Bisect the semicircle *d e* in I, join E I[1], and produce the line to G; E I represents the seat of the common rafter, extending from the wall plate to the junction of the hip-rafters, and the length of the rafter over the seat will of course be equal to B *a* or C *a*. Make I G, therefore, equal to B *a* or C *a*, and join A G, D G. The triangle A G D will then show the extent of the covering of the hip A E D, and A G, D G give the lengths of the hip-rafters.

Similarly, B E and E C represent the seats of the last entire common rafters on the flanks of the roof. Produce B C to F and H, and make B F, C H each equal to the length of the common rafter C *a* or B *a*, and join A F, D H. The triangle A F B is the covering of the triangle A E B, and D H C of D E C.

To find the length and bevel of any jack-rafter on the hip A E D:—From its seat on the plan, shown in dotted lines *g o*, *h p*, draw the parallel lines *g k*, *hl* to the line of the hip-rafter A G in the development; *hl* shows the length of the jack-rafter, and *k l* the bevel of the jack-rafter on the hip-rafter. Any of the other jack-rafters are found in the same way, as shown by the dotted and shaded lines. The down bevel of all the jack-rafters is the angle C *a* E.

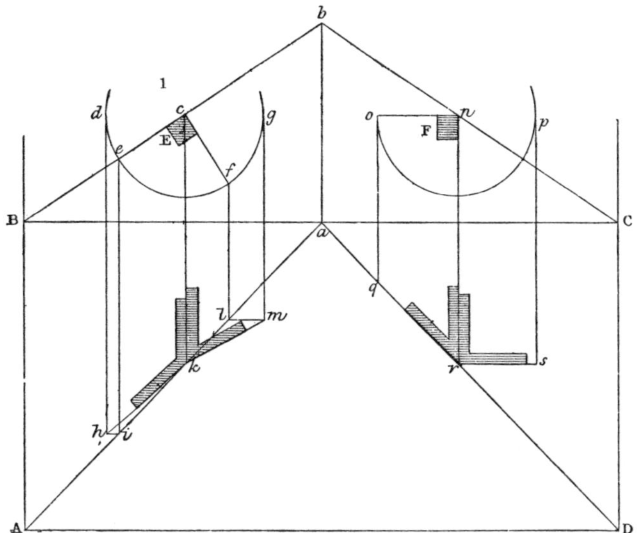

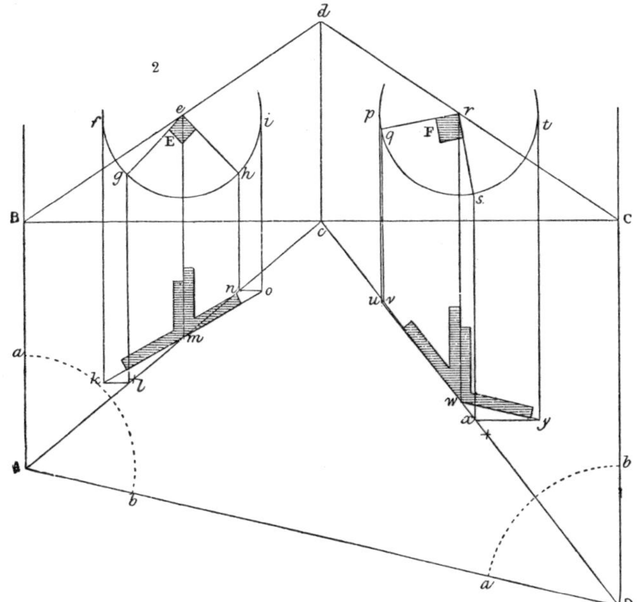

Figs. 718 and 719.—Methods of finding the Bevels on Shoulders of Purlins

Fig. 720, *No.* 2, *shows the method of finding the length and bevel of any single jack-rafter.*

Let A B C be the plan of one angle of a hip-roof, and B E the seat of the hip-rafter. It is required to find the length and bevel of the jack-rafter over the seat D *k e*.

At the point *b*, where the longest side of the jack-rafter meets the hip-rafter, draw *b d* at right angles to the side of the jack-rafter *b e*, and make the angle *b e d* equal to the slope of the roof; draw also *b c* at right angles to the side of the hip-rafter *a b*, and make it equal to *b d*. Produce *b e*, *h k* indefinitely, make *b g* equal to *d e*, and draw *g f* parallel to *e k*; *h b*, *f g* gives the length of the jack-rafter.

[1] Or, more simply, from E draw E I perpendicular to A D.

To obtain the bevel at hb, draw al at right angles to B C and make it equal to kf or eg, and join bl; then the angle lbg is the bevel of the jack-rafter.

To find the bevel of a hip-rafter made by a vertical plane parallel to the planes of the

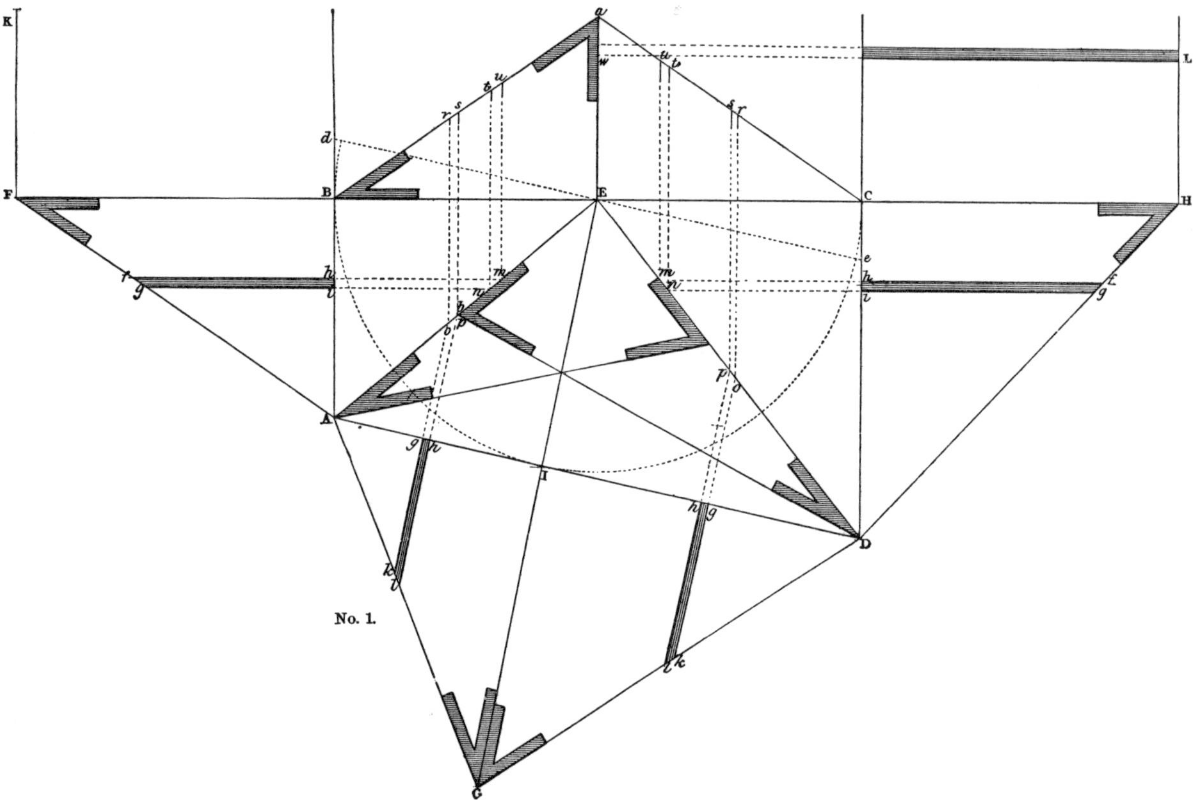

common rafters.

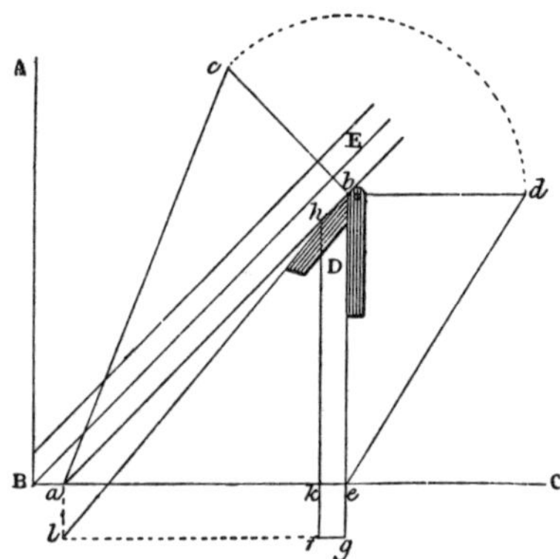

Fig. 720.—Methods of finding the Length and Bevel of Jack-rafters

Let A (fig. 721) be the seat of the foot of the hip-rafter, B the horizontal projection of its upper end, lm the projection of the bevel there, and C the vertical projection of the rafter on a plane parallel to its plane. The backing of the rafter bad at A, and its cross section $ruqvs$ at F, will be evident without description. The method of finding the bevel at gkEh is as follows:—From g draw the horizontal line gi, and from g as a centre, with the radius gh, describe the arc hi. From i let fall a perpendicular in to the horizontal projection meeting Bm produced in n, then draw lno, and plo is the proper bevel.

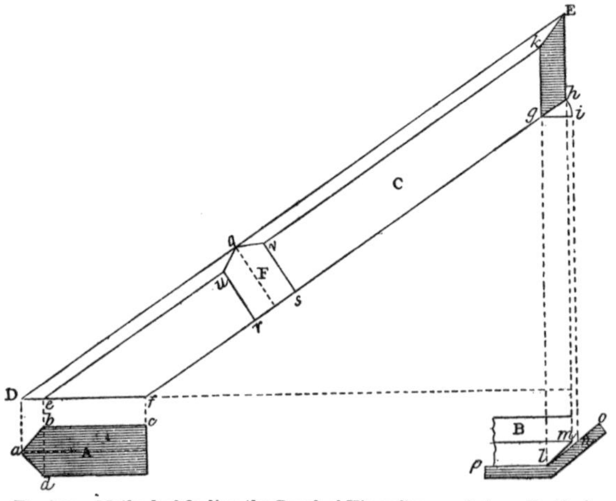

Fig. 721.—Method of finding the Bevel of Hip-rafters made by a Vertical Plane parallel to the Common Rafters

The methods of describing pendentives and domes will be given in the next chapter.

CHAPTER VI

DESIGN AND CONSTRUCTION OF ROOFS

1. ROOF SLOPES AND COVERINGS

Roof Slopes.—"Pitch" roofs may be described as roofs which have an inclination of 18° or more with a horizontal plane. Roofs with smaller inclinations are known as "flat" roofs.

The slope of a roof may be expressed in various ways, namely:—

1. By the angle which the slope makes with a horizontal plane; as, say, 45°.
2. By the angle which the two slopes of a double roof make at the ridge; as 90°.
3. By the amount of rise per unit of half-span; as 1 in 1.
4. By the proportion between the total height of the roof and the width of the span; as "half-pitch".

The four examples just given refer to the same slope of roof; a roof with a slope of 45° may be described as having an angle of 90° at the ridge, or as having slopes rising 1 foot in 1 foot, or as being "half-pitch".

The fourth mode of expression is often adopted, but the relation between the height and span cannot always be expressed by a convenient fraction. The angular measurement, although the least confusing, is not very convenient for the carpenter, so that the third mode of expression will probably be found most generally useful.

In the following table some of the most common roof slopes are expressed in the four different ways.

TABLE VII.—SLOPES OF ROOFS

1. Angle with horizontal ...	18°26′	21°48′	26°34′	30°	33°41′	39°49′	45°	53° 8′	56° 19′	60°	63° 26′
2. Angle at ridge	143° 8′	136°24′	126°52′	120°	112°38	100°22′	90°	73° 44′	67° 22′	60°	53° 8′
3. Rise of each slope per foot	4 ins.	$4\frac{4}{5}$ ins.	6 ins.	($6\frac{13}{14}$ ins.)	8 ins.	10 ins.	12 ins.	16 ins.	18 ins.	($20\frac{7}{9}$ ins.)	24 ins.
4. Pitch $=\frac{\text{height}}{\text{span}}$	$\frac{1}{6}$	$\frac{1}{5}$	$\frac{1}{4}$	($\frac{2}{7}$)	$\frac{1}{3}$	$\frac{5}{12}$	$\frac{1}{2}$	$\frac{2}{3}$	$\frac{3}{4}$	($\frac{13}{15}$)	1

NOTE.—The figures in brackets are very nearly but not absolutely correct.

The slope given to a roof serves the purpose of throwing off more or less rapidly the water of rain or of snow, in order that the materials of the roof may be quickly dry, and that water may not pass easily through the joints of the roof covering.

Some writers have proposed to regulate the slope according to the climate or degree of latitude in which the building is situate, but such proposals are of merely academic interest. In this country, during what may be termed the organic period of architecture, we have the steep-pitched roofs of the Early English or First Pointed style and the nearly flat roofs of the Perpendicular or Third Pointed. It is much nearer the truth to say with Rondelet that the slope given to roofs is dependent on taste alone, with such restrictions only as the more or less perfect nature of the materials imposes. Roofs of quick slope are, however, generally considered most suitable on practical grounds for localities where there is much snow or rain, and roofs of low pitch with widely-projecting eaves and gables for hot sunny climates.

Except in strictly utilitarian structures the pitch of roofs will as a rule be governed by taste and style, rather than by climate and materials. No one would dream of covering a church, erected in the Greek style, with a high-pitched roof like that of a thirteenth-century Gothic cathedral. But, although taste and style are probably the governing factors, practical considerations ought never to be forgotten, and it will be well therefore to note briefly the advantages and disadvantages of slopes great and small. It is obvious that rain and snow will pass from a high-pitched roof more quickly than from one of low pitch; this is not always an unmixed blessing, as large masses of snow may shoot over the eaves-trough and fall upon the heads of persons below. In a high-pitched roof there is less likelihood of rain being driven through the joints of the slates or tiles, and a smaller lap can therefore be given to these. Such a roof, however, not only exposes a greater area to the action of the wind, but also has to bear a greater pressure per square foot, and consequently in many cases exerts a greater thrust on the walls. It is also more expensive, because there is more of it, but when the space in the roof can be utilized, either for separate rooms or for increasing the air-space in the rooms below, the extra expense is often more than counterbalanced by the extra accommodation. A high-pitched roof is difficult to repair, particularly where there is no parapet, and, if lead is used as the covering, the lead will gradually creep down the slopes and either open at the joints or crack.

Table VIII.—ROOF COVERINGS—SMALLEST INCLINATIONS, WEIGHTS, AND WIND-PRESSURE

Materials.	Minimum Rise of each Slope per ft.	Weight per sq. ft.	Wind-pressure per sq. ft.
	ins.	lbs.	lbs.
Flat Roofs.			
Copper	1	1 to $1\frac{1}{2}$	$6\frac{3}{4}$
Zinc	2	$1\frac{1}{4}$,, 2	$13\frac{1}{2}$
Corrugated Iron	2	$2\frac{1}{2}$,, $3\frac{1}{4}$	$13\frac{1}{2}$
Lead	2	$5\frac{1}{2}$,, $8\frac{1}{2}$	$13\frac{1}{2}$
Pitch-roofs.			
Welsh Slates (36 inches to 30 inches)	5	8 ,, 11	26
,, (26 ,, 10 ,, *Firsts*) ...	6	$6\frac{1}{2}$,, 8	30
,, (,, ,, ,, ,, *Seconds*) ...	6	$7\frac{1}{2}$,, 9	30
Westmoreland Slates, *Firsts*	6	9 ,, 11	30
,, ,, *Seconds*	6	$10\frac{1}{2}$,, $12\frac{1}{2}$	30
,, ,, *Thirds*	6	$12\frac{1}{2}$,, $14\frac{1}{2}$	30
Stone Slates	7	24 ,, 30	32
Pan Tiles	6	12 ,, 15	30
Plain Tiles	7	18 ,, 24	32
Wood Shingles	7	3 ,, 5	32
Thatch	12	6 ,, 7	38
Roof-boarding 1 inch thick	...	3 ,, $3\frac{1}{2}$	...
Lath-and-plaster ceiling (exclusive of joists) ...	...	$7\frac{1}{2}$,, 9	...

Note.—The weight of each kind of covering varies according to the thickness, amount of lap, &c. Stone slates, for example, may range from $\frac{3}{4}$ inch to $1\frac{1}{4}$ inch in thickness, and may be laid with laps of 3 to 5 or 6 inches.

In exposed situations roofs may with advantage overhang both at the eaves and gables, as this arrangement keeps the walls drier; in such situations also the roofs ought to be as simple as possible, as valleys, gutters, and hoppers increase the danger of leaks.

"Flat" roofs covered with lead are often used, but are difficult to keep water-tight if the slope is much less than 15°; they have the advantage of requiring only a simple wooden framework, which does not exert any outward thrust on the walls. Copper is well adapted for these roofs, and is lighter than lead as the metal can be used in thinner sheets. Zinc is cheaper, but is not suitable for the acid-laden air of towns.

Roof Coverings.—It is probable that the slopes of roofs were regulated originally according to the materials used as a covering. A thatch of leaves, bark, straw, or reeds, probably the first kind of covering employed, required a steep slope that the water might be speedily thrown off. And when in course of time the more perfect covering of tiles and slates came to be applied, the habit of imitation would for a while prevent any change in the accustomed slope. However this may be, it is evident from the variety of slopes in the same localities that no precise rule can be drawn from existing examples.

For practical purposes what is required is a statement of the lowest slopes which can safely be given to roofs for different kinds of covering materials. This will be found in Table VIII, together with the approximate weight of the coverings per square foot and the wind-pressure.

In calculating the scantlings of roof timbers an allowance must also be made for the weight of the wooden framework. The weight of snow is also sometimes taken into consideration, but it is not a formidable item in this country. Roughly speaking, fresh snow is about one-twelfth the weight of an equal volume of water, so that a cubic foot of fresh snow weighs about $5\frac{1}{4}$ lbs. An allowance of 10 lbs. per square foot will as a rule be sufficient for flat or nearly flat roofs, while for pitch-roofs the allowance may be gradually reduced as the slope is increased; 5 lbs. per square foot will suffice for slopes of 25° to 30°, and a reduction of 1 lb. per square foot may be made for every additional 10°. In calculating the strength of steep roofs, therefore, the weight of the snow may be neglected, especially as it is not likely to act on the roof at a time of maximum wind-pressure.

2. COMMON RAFTERS AND COUPLE ROOFS

Common Rafters.—Before discussing the different types of roofs it will be convenient to give some particulars about common rafters or "spars", as these are required in almost every case. In this country they are commonly fixed 12 inches apart, or 14 or 15 inches from centre to centre, but it is often more economical, especially in roofs without purlins, to adopt a wider spacing. In America the spacing is generally from 16 to 20 inches, but in some localities the rafters are fixed 2 feet and even 3 feet from centre to centre. In Boston the usual spacing for dwellings is 20 inches, but the attic lathing is nailed to furring strips, fixed transversely under the rafters.

The scantling of a common rafter must, of course, be regulated by the load, the length of the rafter between the supports, and the kind of timber used, and, these being known, it may be calculated from the formulas already given (pp. 317–8, Vol. I). It is advisable to measure the length along the slope and not horizontally, as the wind, which often furnishes the greatest portion of the load, acts normally to the slope. The following table gives the scantlings of red deal or fir, white deal, and spruce, required for various lengths of rafters for roofs of quarter-pitch, calculated from the following data:—load (including wind-pressure and boarding, see Table VIII) 42 lbs. per square foot, rafters spaced 15 inches from centre to centre, coefficient of rupture 4200, factor of safety 6.

TABLE IX.—SCANTLINGS OF COMMON RAFTERS

Clear span … … …	4 feet	6 feet	8 feet	10 feet	12 feet	14 feet	16 feet	18 feet	20 feet
Depth, when breadth is 2 ins.	$2\frac{3}{8}$ ins.	$3\frac{1}{2}$ ins.	$4\frac{3}{4}$ ins.	$5\frac{7}{8}$ ins.	7 ins.	$8\frac{1}{8}$ ins.	$9\frac{3}{8}$ ins.	$10\frac{1}{2}$ ins.	$11\frac{5}{8}$ ins.
,, ,, ,, 3 ins.	2 ins.	$2\frac{7}{8}$ ins.	$3\frac{7}{8}$ ins.	$4\frac{3}{4}$ ins.	$5\frac{3}{4}$ ins.	$6\frac{3}{4}$ ins.	$7\frac{5}{8}$ ins.	$8\frac{5}{8}$ ins.	$9\frac{1}{2}$ ins.

From this table the following safe and simple rule, applicable to quarter-pitch roofs and to the kinds of wood generally used for common rafters, can be evolved:—

Let the breadth of the common rafter be 3 inches, then the depth must be half as many inches as there are feet between the supports of the rafter.

It is an easy matter to calculate the depth for rafters of other breadths, as in all rafters of the same length $b d^2$ must be the same. Let it be required to find the depth x of a rafter $2\frac{1}{2}$ inches broad and 16 feet long. A rafter 3 inches broad must be (according to the empirical rule) 8 inches deep. Therefore,

$$3 \times 8^2 = 2\tfrac{1}{2} \times x^2,$$

$$x^2 = \frac{3 \times 8 \times 8}{2\frac{1}{2}} = 76{\cdot}8,$$

$$\text{and } x = \sqrt{76{\cdot}8} = 8{\cdot}76, \text{ or about } 8\tfrac{3}{4} \text{ inches.}$$

In America 2-inch by 6-inch rafters are often used for lengths up to 10 or 12 feet, 2-inch by 8-inch for lengths of 16 feet, and 2-inch by 10-inch for lengths of 20 feet. It is usually more economical to reduce the span of long rafters to 10 or 12 feet by means of purlins or partitions.

In many cases rafters are continuous over a number of supports, and may therefore be of smaller scantlings than stated in the table (see Chapter V of Part II, Section VI).

Fig. 722 shows the common construction of a weaving-shed roof. The rafters rest at the foot on a $3\frac{1}{2}$-inch × 2-inch plate bolted to the flange of the cast-iron gutter, and at the head are notched into a ridge-piece, which is supported from the other gutter by cast-iron standards of T section. These standards are from 6 to 8 feet apart, and the top-lights forming the quicker slope of the roof are secured to them by $\frac{1}{4}$-inch bolts or screws. Plates are bolted under the soffits of the gutters to receive the plasterer's laths.

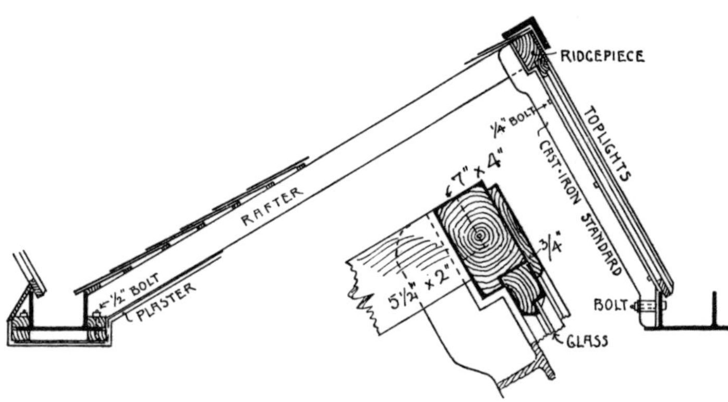

Fig. 722.—Section of Weaving-Shed Roof and Detail at Ridge

Couple Roofs (fig. 723, No. 1).—The simplest roof of two slopes is the couple roof, consisting of pairs of common rafters abutting against a ridge-piece at the top and with the feet resting on the walls or nailed to wall-plates. The rafters obviously exert an outward thrust on the walls, and this form of roof is usually considered suitable for small spans only, up to about 10 or 12 feet; much depends, however, upon the weight of the roof covering and the resistance of the walls. If the walls and wall-plates are tied together by (say) the attic floor-beams, spans up to 30 feet may be roofed in this way. The ends of the rafters may be cut to the required bevels, and simply nailed to the ridge and wall-

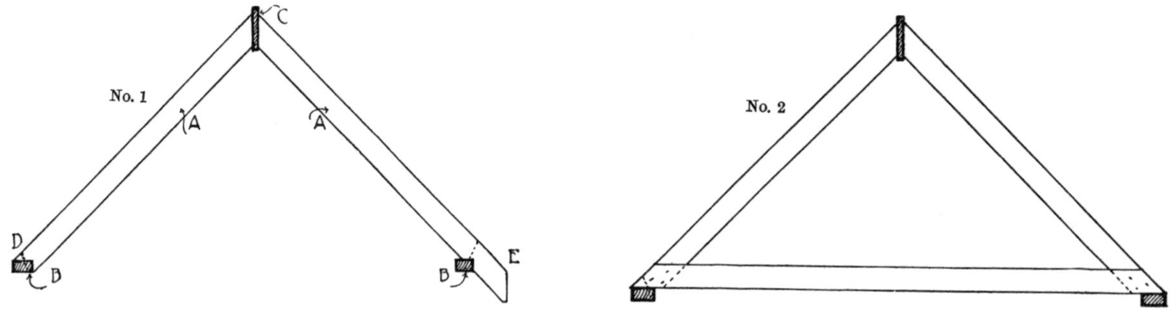

Fig. 723.—No. 1, Couple Roof: A A, Rafters; B B, wall-plates; C, ridge-piece; No. 2, Couple-close Roof

plates, or the feet of the rafters may be cut out as shown at D or E; the best method is that shown at D.

Couple-close Roofs (fig. 723, No. 2).—These are similar to couple roofs, but with an additional member inserted to connect the feet of each pair of rafters in order to prevent the outward thrust on the walls. The lower members serve the purpose of tie-beams, and can also be utilized as ceiling joists. In small roofs where flat ceilings are not required sufficient resistance to the thrust can be obtained by fixing a tie from wall-plate to wall-plate every 5 or 6 feet; the ties may be simply nailed for rough work, or secured with cogged joints in addition to nails.

This kind of roof is generally considered suitable for small spans only, as the rafters are not supported throughout their length. Of course, as already pointed out, the rafters can be of larger size for larger spans, and in many cases the cost of the additional timber will be more than counterbalanced by the saving in labour which is effected on account of the simplicity of the construction. That system of construction which is most economical in material, and which is therefore theoretically the best, is not always the best from a practical point of view, as it may involve an extra amount of labour altogether disproportionate to the saving in material. The scantling of the rafters can be easily calculated from the formulas already given, or may be copied from Table IX.

3. TRUSSED-RAFTER ROOFS

Trussed-rafter roofs are more or less elaborate developments of the couple-close roof. Each pair of rafters is formed into a truss, which may be so designed as to exert little or no outward thrust on the walls. In many cases also the rafters are supported by the other members at one or more points in their length. In these roofs the weight is distributed equally along the wall, instead of being concentrated at a series of distant points as in the case of roofs where purlins, supported by trussed principals, are used. The principal varieties are (1) the collar roof, (2) the double-collar, or collar-and-tie, (3) the scissor-beam, and (4) the compound.

Collar Roofs.—It is often said that rafters must not be more than 8 feet long between the supports, and that, if the length from wall-plate to ridge exceeds 8 feet, additional members are required. For low-pitch roofs a single cross-piece is generally inserted horizontally to connect each pair of rafters. This cross-piece is known as a collar, and is often said to serve two purposes, (1) to resist the outward thrust of the rafters, and (2) to stiffen the rafters and prevent them sagging. In other words, it must serve both as a tie and as a strut at the same time. Under the dead load of the roof covering this is impossible, as the reader can demonstrate in a few minutes by means of a model consisting of two cardboard rafters pinned together at the apex by a single pin, and united at about half their height by a piece of string to serve as a collar. If weights are suspended from any points in one or both rafters, and the feet of the rafters are free to spread, it will be found that the string is always in tension, and the strength of the truss depends on the rigidity of the rafters. The stresses in trusses of this kind have already been considered in Chapter VIII, Section VI, Case 12, and it will be sufficient to point out here that in ordinary circumstances the collar serves as a tie only and does not increase the strength of the rafters. When, however, the feet are prevented from spreading by means of parapets or other suitable abutments, the collar serves as a strut and the scantling of the rafters can be reduced.

Ordinarily, however, the collar is a tie, and its position becomes a matter of importance. The nearer it is placed to the feet of the rafters, the better will it fulfil its duty, and the less danger will there be of the walls being thrust out. In any case it ought not to be placed above the middle of the height of the rafters. In all roofs of this class there is a certain amount of settlement owing to the shrinkage of the wood, defective workmanship, and other causes, and the higher the collar is placed the greater will be the spread at the feet of the rafters.

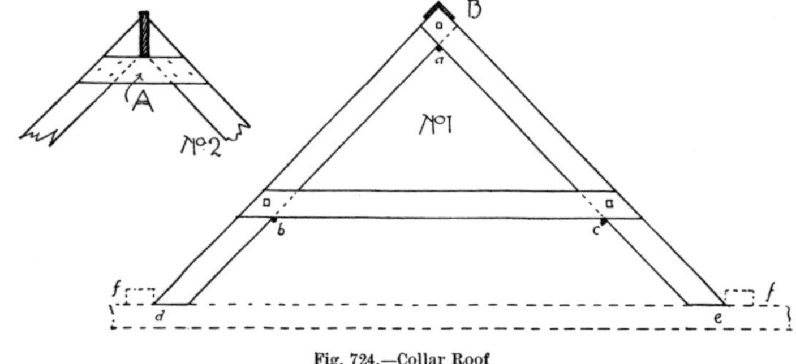

Fig. 724.—Collar Roof

An ordinary type of collar roof is shown in fig. 724. The joints for the ends of such collars have already been discussed in Chapter I, where it was pointed out that dovetailed halving cannot be depended upon, on account of the probable shrinkage of the wood. The halving also reduces the strength of the rafters at or near the point of greatest stress. A wrought-iron bolt with nut and washers forms a simple and secure fastening; a diameter of half an inch is sufficient for nearly all spans which are likely to be adopted. For rough work clenched nails may be used. The rafters may be bevelled at the top and spiked to the ridge-piece, additional strength being given if necessary by a light cross-piece nailed to the rafters immediately below the ridge, as shown at A in No. 2. Another method consists in halving the heads of the rafters, and bolting them together as shown at B; two boards about 5 inches by $\frac{3}{4}$ inch nailed to the rafters take the place of the ridge-piece. The bolt-holes ought always to be slightly less in diameter than the bolts, so that these will fit tightly when driven through.

The latter method of construction requires less labour in making and fixing. Set out on a wood floor, or on a piece of level ground, a full-size outline of one couple or truss. Fix at the level of the feet of the rafters a piece of wood *d e* (fig. 724), with blocks at *f f* to give the full width of the couple, and insert pegs at *a*, *b*, and *c*. Cut the feet of the rafters to the required bevel, and then lay them to touch the pegs *b a* and *c a* respectively; the heads can then be marked and cut to form the apex, and the hole bored and bolt inserted. Then lay the collar to touch the pegs *b* and *c*, cut the ends to the required bevel, bore the holes and insert the bolts. The washers and nuts can then be fixed, and the couple is ready for fixing.

One objection to the collar roof is the difficulty of adapting it to irregular plans, and to roofs with dormers or transverse gables. In applying it to a building wider at one end than the other, all the couples will be different in size. The collars, however, ought to be

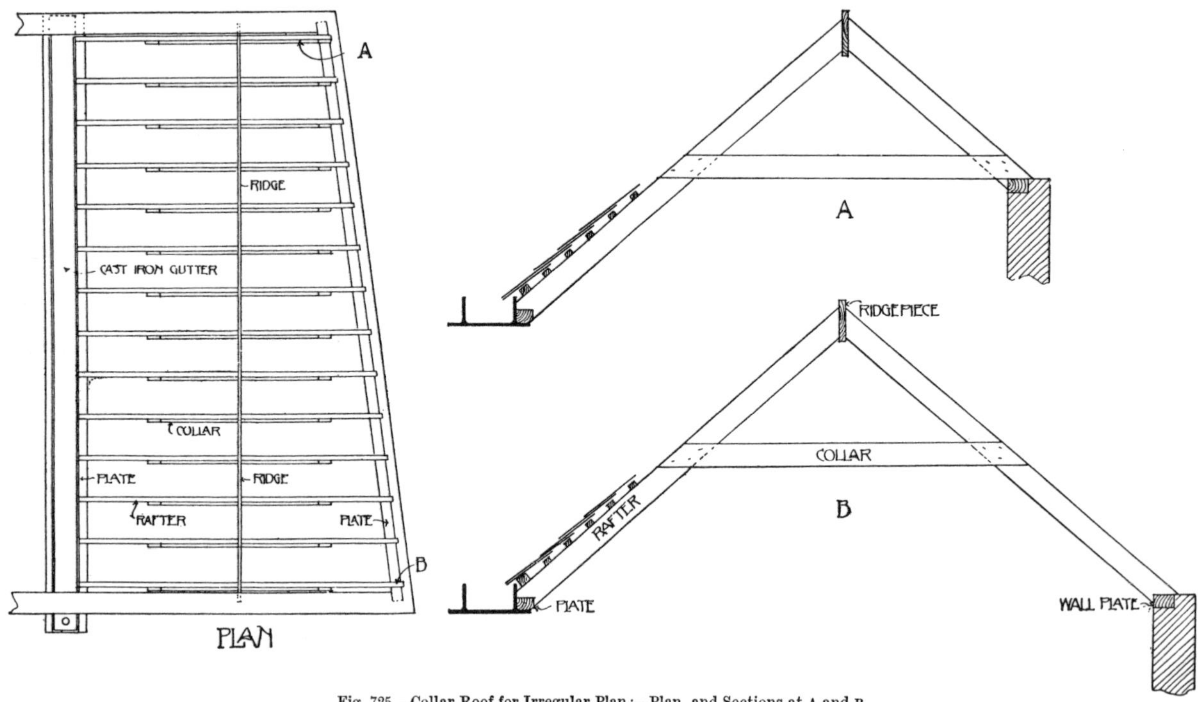

Fig. 725.—Collar Roof for Irregular Plan:—Plan, and Sections at A and B

all on the same level. If both eaves are level and the ridge rises from the narrow to the wide end, every couple will be essentially different. The simplest method is to make one eaves level, and to make the ridge also level and parallel to the level eaves, and to let the other eaves slope as required by the amount of skew and the pitch of the roof. The latter arrangement is shown in fig. 725. It will be seen that the only difference in the couples is the length of the rafters below the collars on the right-hand side of the roof.

Small gables may be trimmed for, as shown in fig. 726. The trimming couples must of course be stronger than the others, and two bolts may with advantage be used at the end of the collars, these being made of greater depth for the purpose; or two collars may be used, so that the single bolt is brought into double shear. The trimming couples ought also to be placed so that the gable wall serves as an abutment.

Sometimes a collar is fixed to certain pairs of rafters only, say every fifth or sixth, and purlins are inserted to support the intermediate rafters, as shown in fig. 727. The coupled rafters are therefore heavily loaded, and must be made considerably stronger than the others. They are almost certain to exert an outward thrust on the walls, and this method of construction cannot therefore be recommended except for light roofs.

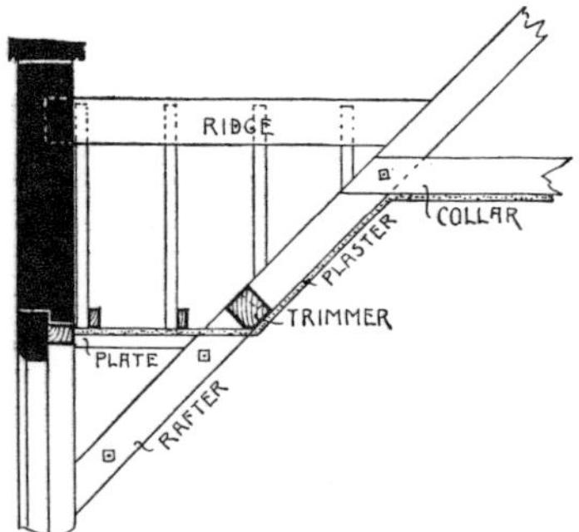

Double-Collar or *Collar-and-Tie Roofs.*—Two tiers of collars may with advantage be used for roofs of lofty pitch, as shown in fig. 728. In this case the upper collar *a b* is in compression, and may be framed between the rafters as shown, and secured with tenons or spikes, and the lower collar *c d* is in tension.

Each rafter is in this case supported by the upper collar, and the scantling may be calculated from the longer of the two portions into which the collar divides it, namely, from the wall-plate to the collar or from the collar to the ridge. The lower collar is often placed at one-third of the height of the roof and the upper at two-thirds, but it is better to fix the upper collar at half the height, and the lower as near the feet of the rafters as possible. If the lower collar has a heavy ceiling to support, it may be necessary to brace it up by means of a king-bolt, or by two wood braces nailed to the collar and rafters (as shown by the dotted lines).

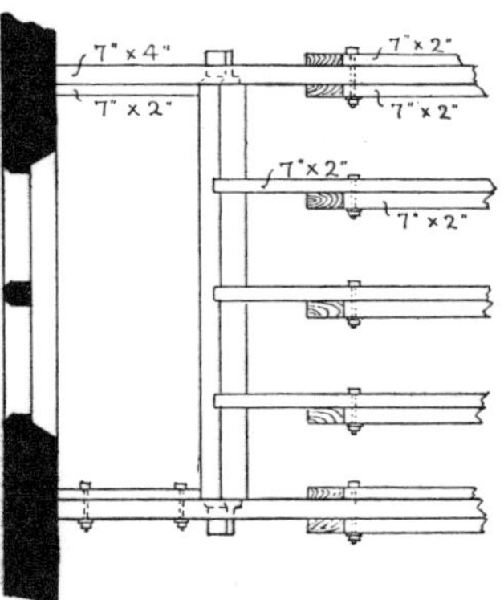

Fig. 726.—Collar Roof trimmed for Gable

The term "double-collar roof" may also be applied to a roof in which the single collar is replaced by two pieces fixed to the rafters on opposite sides but on the same level.

Scissor-beam Roofs.—These take their name from two oblique members which cross each other in the form of a pair of scissors. The stresses in such trusses were considered in Case 13, Chap. IV of Part I, Section VI, and four outline illustrations were there given.

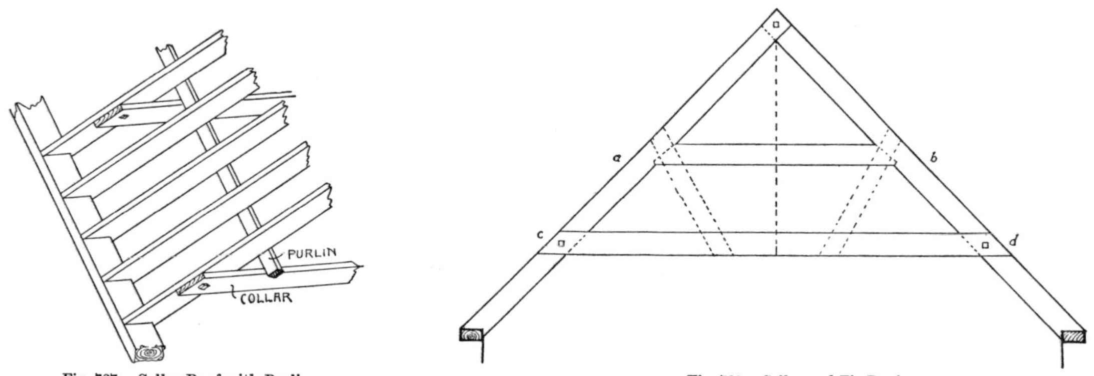

Fig. 727.—Collar Roof with Purlins

Fig. 728.—Collar-and-Tie Roof

As the details of construction do not present any novel features, no further description is necessary.

The German Truss is a modification of the scissor-beam truss. As shown in fig. 729, a king-post and two oblique ties are introduced. The collar is in compression, the rafters are subjected to compression and transverse stress, and the other members are in tension. The simplest method of construction is that shown at A, the oblique members being placed on opposite sides of the other members, but this introduces a slight torsional stress in the truss. To avoid this the oblique members may be double, as shown at B, the collar and king-post being halved together, or the king-post as well as the oblique members may be double, as shown at C. The rafters being supported in the middle are more than twice as strong as in a couple-close roof of the same span. The ends of the collars may be halved on to the rafters and secured with nails or bolts. A board may be clinch-nailed to unite the three pieces at the apex. Trussed-rafter roofs of this and other kinds involving a considerable amount of labour may, for the sake of economy, be spaced farther apart than ordinary rafters, stronger laths being used for the slating, and furring strips being fixed to receive the plasterer's laths.

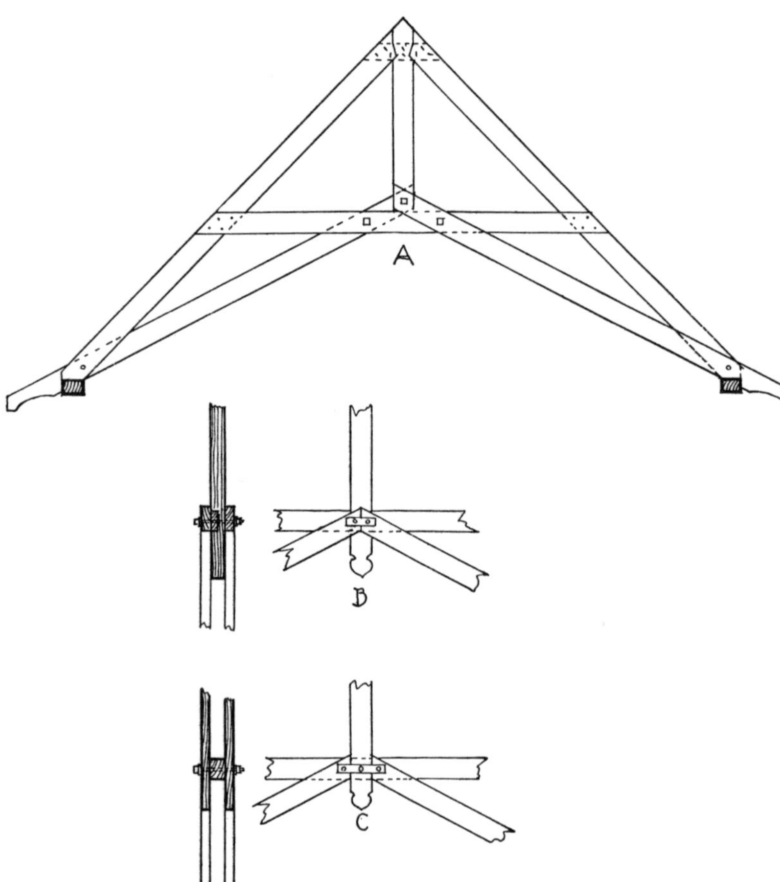

Fig. 729.—The German Truss

Mansard roofs for houses may be constructed entirely with trussed common rafters, as shown in fig. 730. The joists of the attic floor serve as the main ties, and are spiked to the wall-plates. In No. 1 the common rafters forming the lower slopes of the roof are nailed to the joists, and supported in the middle by ashlaring. They are cut out at the top in bird's-mouth form to support the continuous plate, on which the upper rafters and ceiling joists rest. A more elaborate arrangement is shown in No. 2, where the lower slopes of the roof are curved, and an eaves cornice of wide projection is constructed. A partition is introduced at A, so that the walls of the rooms are vertical.

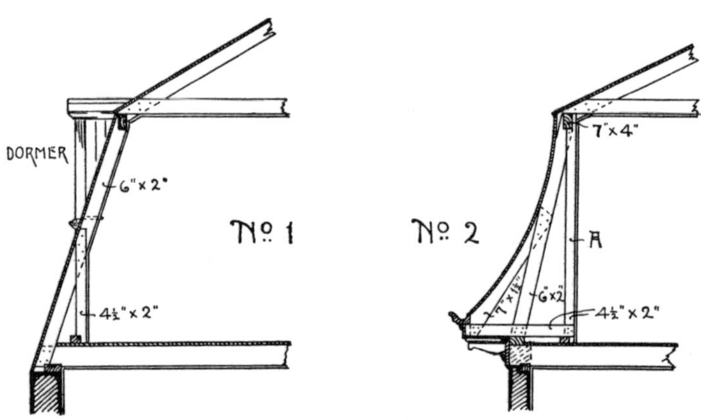

Fig. 730.—Mansard Roofs

No. 1, with Straight Slopes; No. 2, with Curved Lower Slope and Wide Cornice

The roof of the house illustrated in Plates XIX to XXIV is an example of a trussed-rafter roof. The rafters of the main roof are supported from the floor joists by means of ashlaring or partitions at a short distance from the eaves, and by 4½-inch by 2-inch collars at a higher level, the collars forming ceiling joists. The roof over the two-storied part of the building has trussed rafters with ties (forming ceiling joists) near the feet, and collars at about half the height of the truss.

Fig. 731 gives elevations and details of a trussed-rafter Mansard roof at Glasgow. No. 1 is a transverse section of the roof, showing the lower rafters framed into the floor joists and supporting the pole-plates at the top. The ceiling joists rest on the pole-plates and on intermediate partitions, and are connected with the rafters of the upper slopes by vertical and raking struts. The longitudinal section (No. 2) shows the stone jambs and

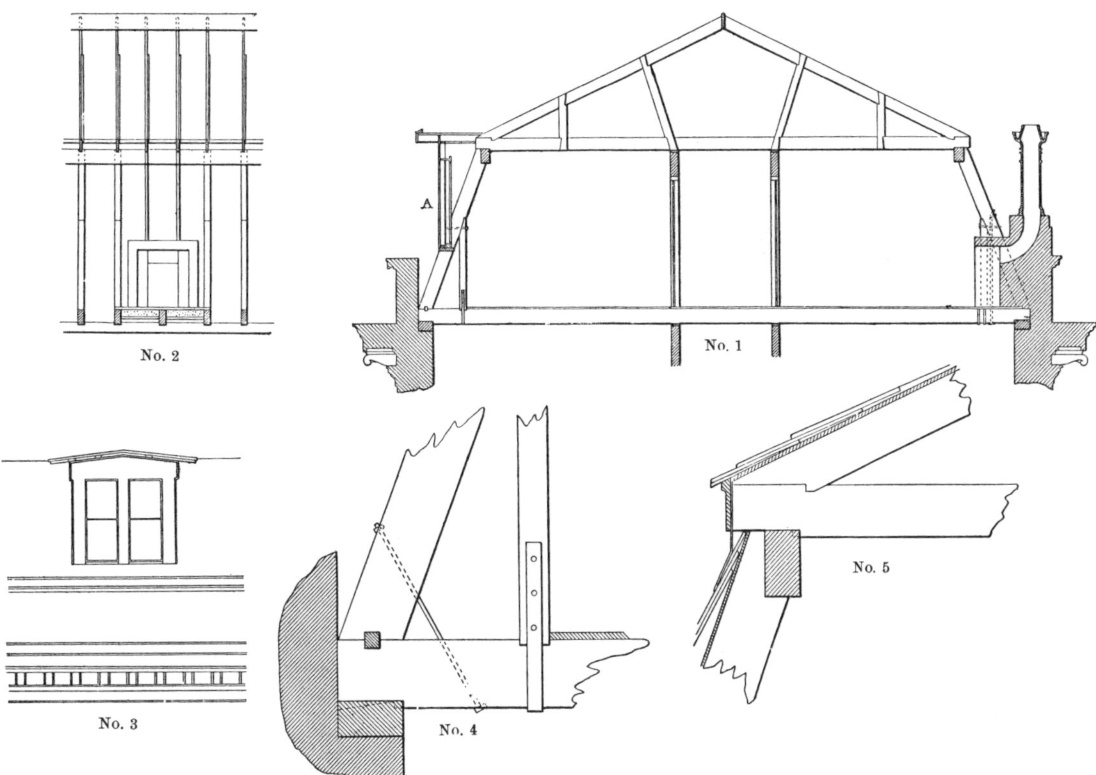

Fig. 731.—Mansard Roof with Dormer, &c.

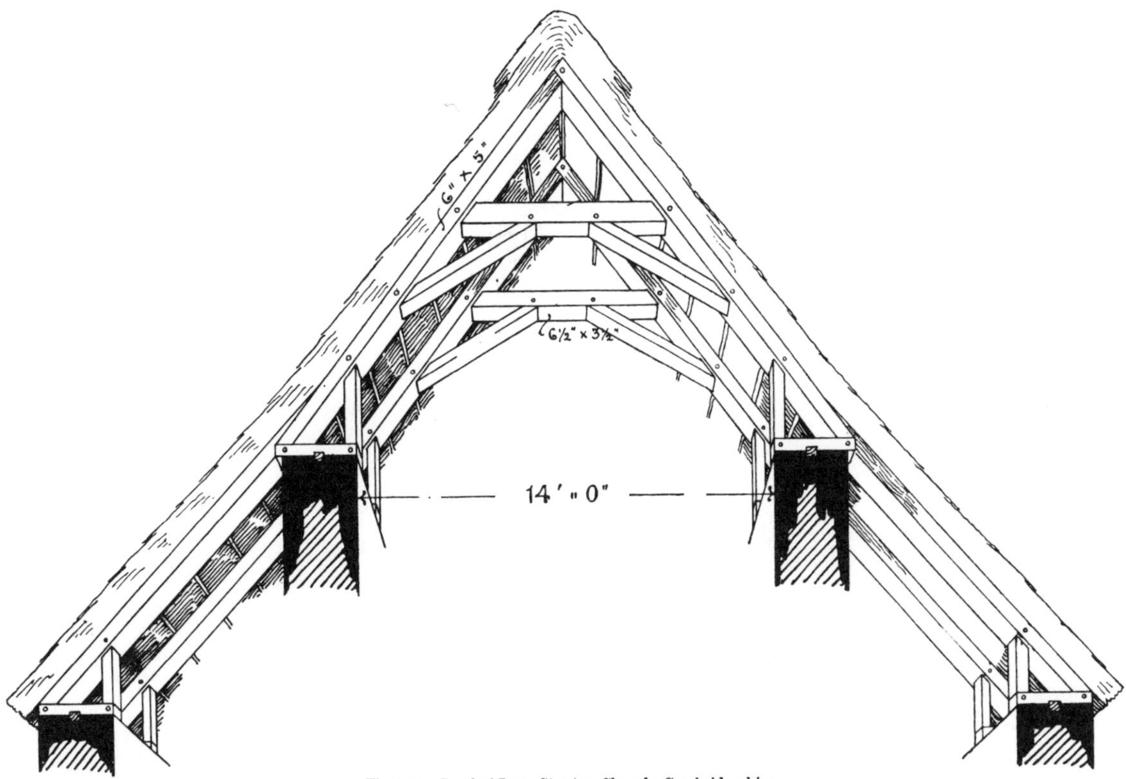

Fig. 732.—Roof of Long Stanton Church, Cambridgeshire

hearth for the stove suspended from the lower rafters. No. 3 is an external elevation of the dormer at A, &c., and Nos. 4 and 5 details of the roof framing.

Compound Trussed-rafter Roofs.—Hitherto we have considered strictly utilitarian structures. Trussed-rafter roofs have, however, been largely used with a view to artistic effect as well as utility, notably in mediæval churches. Additional members are often introduced, partly for strength and partly for effect, and the scantlings are generally larger than is theoretically necessary.

The roof of Long Stanton Church, Cambridgeshire (fig. 732[1]), is a simple example of a collar roof with braces connecting the collar and rafters. The span of the nave is only 14 feet. The angle of the roof slope is 52°, and the cover is thatch. The rafters

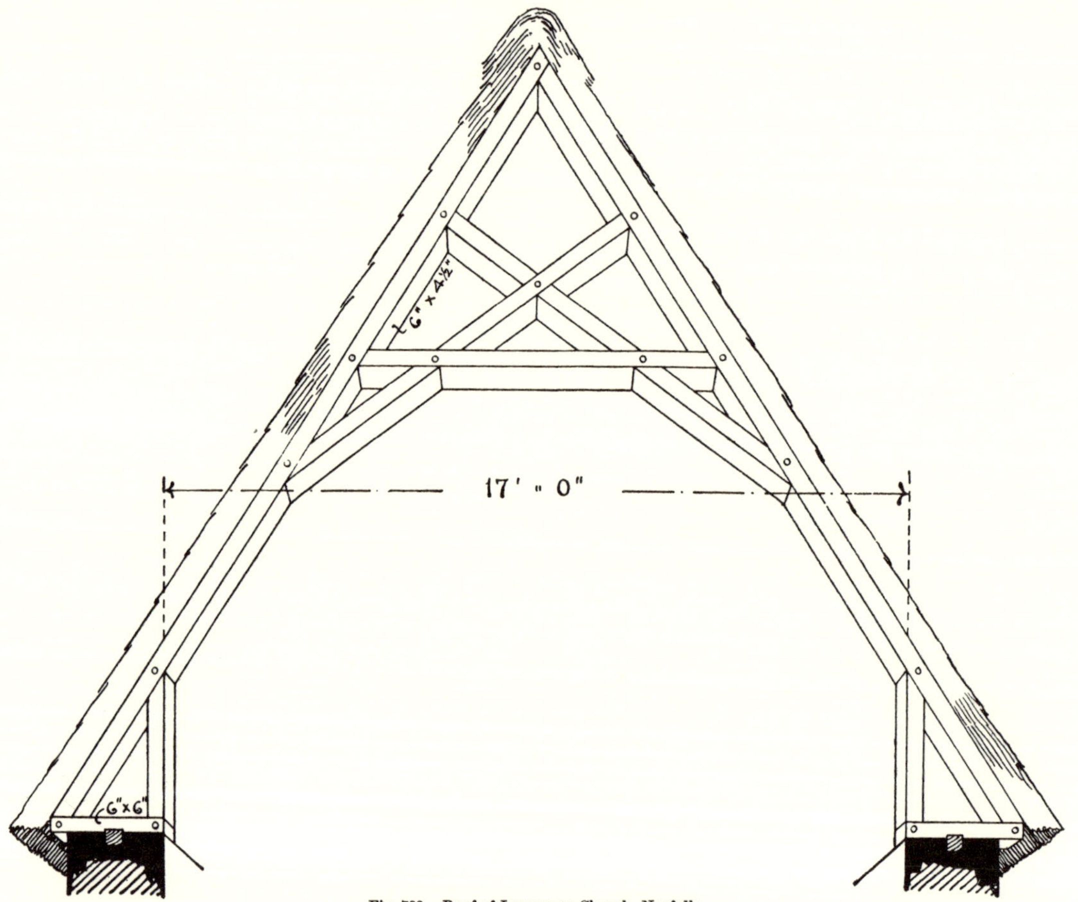

Fig. 733.—Roof of Lympenoe Church, Norfolk

are 5 inches by 6 inches, the collar 3½ inches by 6½ inches, the wall-plates 3 inches by 2½ inches, and the space between the rafters 12 inches. The members are tenoned into each other, and secured with wood pins. The feet of the rafters are firmly supported on ties extending across the wall above the plate, and from the inner ends of these ties vertical studs or struts (known as "ashlaring") are carried up and pinned into the rafters.

The roof over the nave of Stowe Bardolf Church, Norfolk, is of similar design, but with an additional collar midway between the main collar and the apex. The span in this case is 23 feet 10 inches between the walls, and all the members measure 4½ inches by 5 inches; the rafters are fixed 21 inches apart, and the covering is thatch.

The next example (fig. 733) is from Lympenoe Church, Norfolk, and is a modification of the scissor-beam truss, the collar intersecting the scissor-beams. The clear span is 17 feet, and the angle of the slope 57°. The wall-ties measure 6 inches by 6 inches, and the other members 6 inches by 4½ inches, the space between the rafters being 21 inches. The scissor-beams and collars are halved and pinned together at their intersections. The roof of the nave of Ely Cathedral is of very similar design, but on a larger scale.

[1] Figs. 732 to 734 are copied from Brandon's *Open Timber Roofs of the Middle Ages.*

Roofs of these kinds are often boarded underneath to form the ceilings, and the boarding divided into panels by means of moulded ribs, often with carved bosses at the intersections. The example illustrated in fig. 734 is from St. Mary's Church, Wimbotsham,

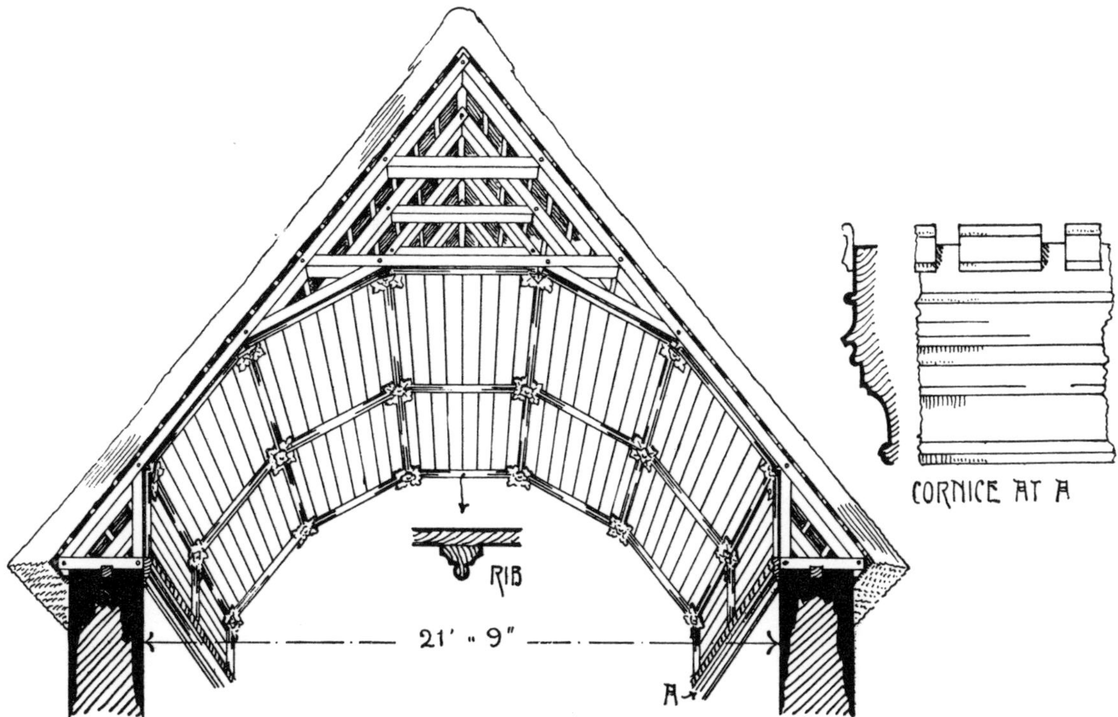

Fig. 734.—Roof of St. Mary's Church, Wimbotsham, Norfolk

Norfolk. The clear span of this is 21 feet 9 inches; the rafters, collars, and other members measure 4½ inches by 4 inches, and the space between the trusses is 21 inches.

Arched braces are sometimes introduced instead of the diagonal braces so as to form a semicircular instead of a polygonal ceiling.

4. ROOF-PRINCIPALS

Advantages.—It is clear that a considerable amount of labour is involved in forming every pair of rafters into a truss, if this is of intricate design; and, when the span approaches 30 feet, the arrangement is often wasteful of material. For spans exceeding 20 feet recourse is now generally had to the trussed-principal roof, in which framed trusses are fixed across the space to be roofed at convenient distances. Between these trusses, and supported by them, horizontal pieces known as *purlins* are placed, and on these the common rafters are supported.

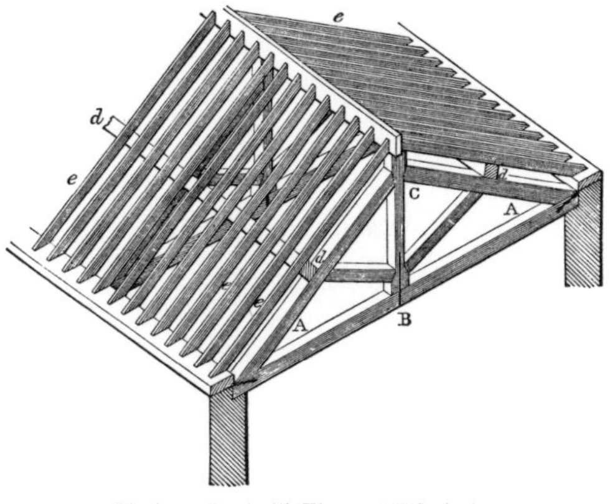

Fig. 735 illustrates this kind of roof, in which A A B C is the trussed frame of carpentry, called a *principal*, *d d* are the purlins, and *e e* the common rafters. The wall-plates and ridge-piece remain as in the roofs already described. The space between two principals is known as a *bay*.

Fig. 735.—Roof with King-post Principals

This system of construction lends itself to the formation of transverse gables, dormers and other projecting features, and the concentration of the weight of the roof at somewhat distant points is often an advantage, as in the case of roofs supported on wood or iron columns or on piers between large windows. It can also be economically used for much larger spans than the trussed-rafter roof.

Width of Bays.—No hard-and-fast rule can be laid down as to the width of the bays into which the roof is divided by the principals. Much depends upon the position of the supports. The general width, however, is from 8 to 12 feet. Sometimes the main principals are fixed at a greater distance, but between each pair an intermediate truss of lighter character is placed; this arrangement is not infrequently seen in churches, the main principals being over the columns or piers of the nave, and the others over the crowns of the arches between the nave and aisles.

Types of Roof-trusses.—The truss shown in fig. 735 is known as the *king-post truss*; the vertical member C B is the king-post, the horizontal member A A is the tie-beam, the outer raking members C A are the principal rafters, and the inner raking members are the struts. The king-post and tie-beam are in tension, and their places may be taken by wrought-iron rods; the other members are in compression.

It is sometimes inconvenient to have the centre of the truss occupied by a member, especially where it is necessary to have apartments in the roof. In such a case recourse may be had to *the queen-post* truss. Two suspending posts, known as queen-posts, are used, and a fourth element is introduced, namely, the *straining-beam,* extending between the heads of the posts. The principle of trussing is the same. The rafters, struts, and straining-beam are compressed, and the tie-beam and queen-posts are in a state of tension. For larger spans three or more suspending posts may be used, and may be placed either vertically or obliquely. Such trusses may be termed compound tie-beam trusses.

Instead of a tie-beam, connecting the *feet* of the principal rafters, a collar-beam is sometimes used to connect the rafters at a higher level. This is known as a *collar-beam principal* or *truss.*

In some roofs, for the sake of effect, the tie-beam does not stretch across between the feet of the principals, but is interrupted. In point of fact, although occupying the place of a tie-beam, it does not serve the same purpose. It is called a *hammer-beam.* An outline diagram of one form of hammer-beam truss was illustrated in fig. 556.

There is still another class of principals in which tie-beams are not used. Such are the curved principals of De Lorme and Emy. In the system of Philibert de Lorme, arcs formed of small scantlings of timber are substituted for the framed principals; and in that of Colonel Emy, laminated arcs are used.

Classification.—The principals of roofs may therefore, in respect of their design, be divided broadly into two classes—*First,* Those with tie-beams; and, *Second,* Those without tie-beams.

The *first* class may be further divided into—1. King-post trusses; 2. Queen-post trusses; 3. Compound trusses.

The *second* class may be arranged as follows:—1. Collar-beam roofs; 2. Hammer-beam roofs; 3. Curved-principal roofs.

Hints on Construction.—It may be well, before describing the examples of these roofs which are given in the following pages, to note some practical memoranda of construction which are applicable to nearly every variety.

Compression-members should not be unduly cut into or mortised on one side, so as to cause lateral yielding. They should be as nearly square in cross section as possible, as this form offers a greater resistance than an oblong section of the same area.

Purlins should never be framed into the principal rafters, but should be notched. They should also be in as long pieces as possible, that is to say, continuous over two or more bays.

Rafters laid horizontally are very good in construction, and in many cases cost less than purlins and common rafters.

The ends of *tie-beams* should be kept with a free space round them, to prevent decay.

It is said that girders of oak in the Chateau Roque d'Ondres, and girders of fir in the ancient Benedictine monastery at Bayonne, which had their ends in the wall wrapped round with plates of cork, were found sound, while those not so protected were rotten and worm-eaten. It is an injudicious practice to give an excessive camber to the tie-beam: it should only be drawn up when deflected, as the parts come to their bearings.

A strut should always abut immediately underneath that part of the rafter whereon the purlin lies. The diagonal joints of struts should be left a little open at the inner part, to allow for the shrinkage of the heads and feet of the king and queen posts.

The breadth of the different members of the truss may with advantage be the same, the depth only being varied, as this arrangement facilitates construction and allows the iron straps to be straight from end to end. Cranks or bends in *iron ties* should be avoided as much as possible.

And, as an important final maxim—*Every construction should be a little stronger than strong enough.*

5. ROOF-PRINCIPALS WITH TIE-BEAMS

I. *King-post Trusses.*—It is commonly said that king-post trusses ought not to be used for spans greater than 30 feet. As there is only one strut under each principal rafter, only one purlin can be fixed on each side of the roof without subjecting the rafter to a transverse stress in addition to the compressive stress which it is primarily intended to bear. In a roof of 30 feet span, with a slope of 30°, the distance between this purlin and the wall-plate or ridge will be about 9 feet, and the common rafters must therefore be made of sufficient size to bear (without deflection) the weight of the roof covering over this space. The tie-beam also is only braced up at one point, each unsupported portion being therefore 15 feet long. To avoid these long spans in the common rafters and tie-beams, it is better to use some other form of roof for spans much exceeding 30 feet. Of course two or more purlins may be fixed over each principal rafter of a king-post truss, but the scantling of the rafter must be increased in order to bear the transverse stress and to compensate for the reduction in the compressive resistance of the rafter which the transverse stress involves.

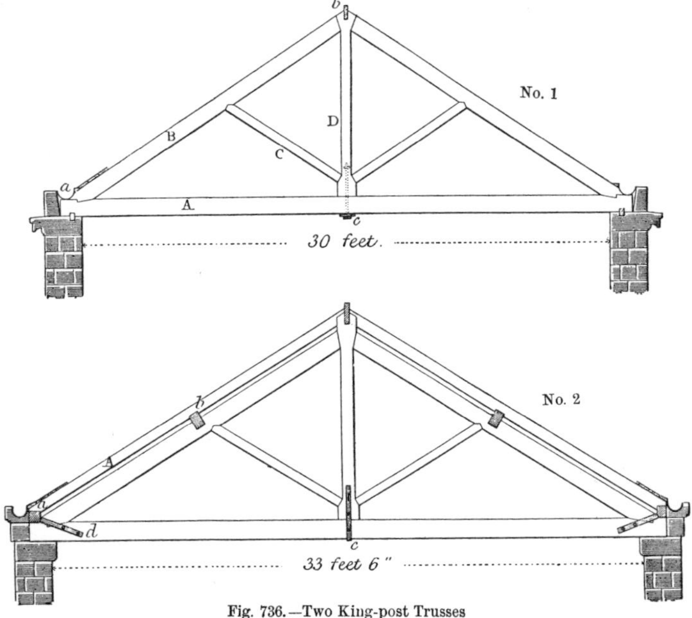

Fig. 736.—Two King-post Trusses

Two examples of king-post trusses are given in fig. 736. No. 1 has a clear span of 30 feet; the tie-beam is 13 × 5 inches, king-post $7\frac{1}{4} \times 5$ inches, principal rafters $8\frac{1}{4} \times 5$ inches, and struts $4 \times 2\frac{1}{2}$ inches. The tie-beam is cogged on to short wall-plates, and is secured to the king-post by a bolt passing up into the latter, the nut being inserted in a mortise cut to receive it. In the second example, the span is 33 feet 6 inches, and the slope 30°. The ends of the tie-beam are shown to rest on through-stones; in some cases dowels are inserted into the tie-beam and through-stones to prevent movement. The wall-plate carrying the feet of the common rafters is shown at *a*; as a rule, the wall is carried up to support this continuously. If, however, the plate is merely supported by the ends of the tie-beams, it becomes a beam, known as a "pole-plate", and must be strong enough to carry the load transmitted by the common rafters in each bay. The purlin *b* is in this case framed into the principal rafter, but it is better to notch or cog the purlin on to the top of the rafter, and to support it by means of a cleat spiked to the rafter, as shown at K in fig. 4, No. 1, Plate XXVII. The king-post is fixed to the tie-beam by a wrought-iron strap or stirrup *c* with gibs and cotters, and the foot of the principal to the tie-beam by a similar stirrup *d* secured with bolts. The tie-beam is $11\frac{1}{2} \times 6$ inches, king-post $7\frac{3}{4} \times 6$ inches, principal rafters 10×5 inches, struts $4 \times 2\frac{1}{2}$ inches, and purlins 10×6 inches.

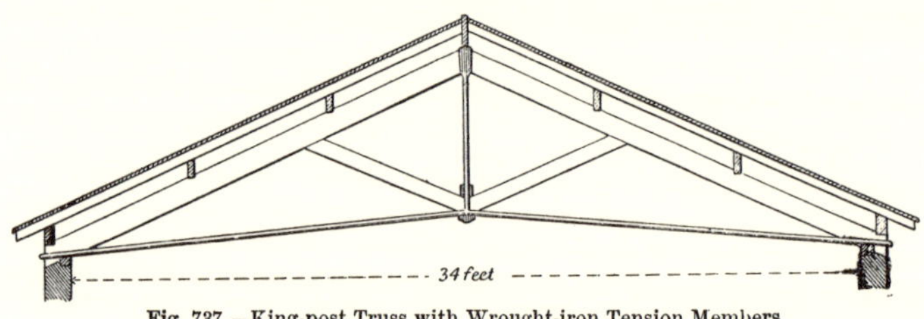

Fig. 737.—King-post Truss with Wrought-iron Tension Members

The truss shown roughly in fig. 737 was used many years ago over a shed at one of the Liverpool docks, and has wrought-iron rods instead of the tie-beam and king-post. The two purlins on each side cause transverse stresses in the principal rafters, and these were consequently made of larger scantling than would otherwise have been necessary. The scantlings are as follows:—Principal rafters, 12×8 inches; struts, 8×8 inches; purlins, 10×4 inches; common rafters, $4\frac{1}{2} \times 2$ inches; tie-rod and king-rod, $1\frac{1}{2}$ inch diameter. The rafters and struts appear to be unduly large, but in the absence of information as to the width of the bays it is impossible to give a definite opinion on the point.

A simple form of king-post truss is shown in fig. 738. It occurs in a shed on the quay at Sundsvall, the well-known Swedish timber port, and was measured by the writer in 1900. The clear span is about 30 feet, and the total width of roof, including overhanging

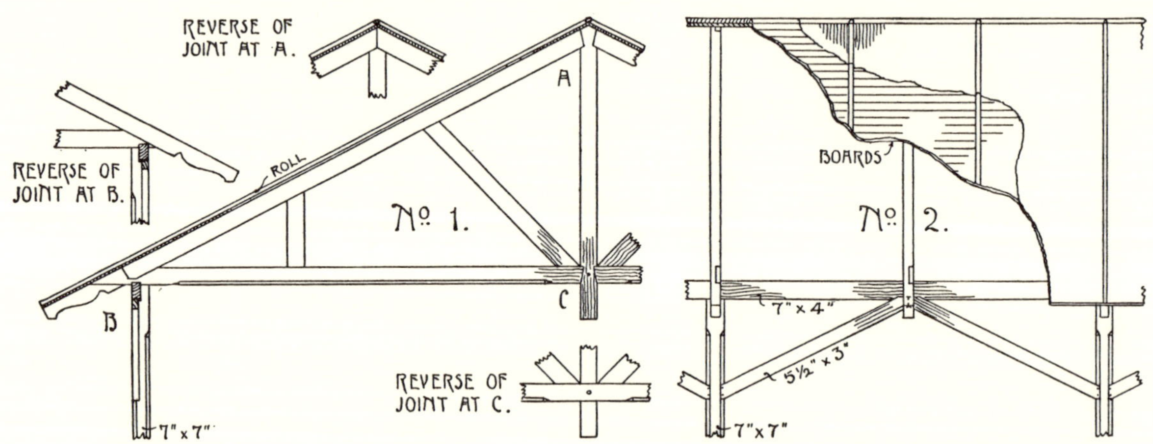

Fig. 738.—Roof over Shed on Quay at Sundsvall, Sweden. No. 1, Half elevation of truss; No. 2, side elevation of one double bay

eaves, about 37 feet. All the members of the truss are of the same scantling—namely, 7×4 inches. The ends of the tie-beam are connected to the principal rafters by dove-tailed halving, the pitch of the dove-tail being much quicker than is common in this country. Halving is also adopted at the head and foot of the king-post. The struts appear to be simply tenoned into the rafters and tie-beam. A short strut is inserted to support each rafter about midway between the joints at the end of the tie-beam and the head of the principal strut. A trenail is inserted in the joint at the foot of the king-post; the other joints may be secured in a similar manner or nailed. There are no purlins, but horizontal boarding is nailed directly to the principal rafters; wood rolls run from the eaves to the ridge about 4 feet apart, and the roof is covered with tarred paper. The roof is supported on wooden posts 7 inches square, and about 13 feet from centre to centre; the intermediate trusses rest on the 7×4-inch pole-plates, which are stiffened by $5\frac{1}{2} \times 3$-inch struts framed into the posts. The shed is in an exposed situation, and in winter is probably covered with snow to a considerable depth.

Tredgold's Rules.—Tredgold's rules for calculating the scantlings of king-post roofs were formulated in the early part of the nineteenth century, and although they are of an empirical nature, they have been so frequently adopted by writers on carpentry, either expressly or tacitly, as to be worthy of recapitulation. They are supposed to be applicable to "Baltic pine" timber of good quality, and to give the safe scantlings for a total load (including wind-pressure) of $66\frac{1}{2}$ lbs. per square foot of surface.

They are as follows:—

1. *To find the dimensions of the principal rafters.*

Multiply the square of the span in feet by the distance between the principals in feet, and divide the product by 60 times the rise in feet: the quotient will be the area of the section of the rafter in inches.

If the rise is one-fourth of the span, multiply the span by the distance between the principals, and divide by 15 for the area of section.

When the distance between the principals is 10 feet, the area of section in inches is two-thirds of the span in feet.

2. *To find the dimensions of the tie-beam, when it has to support a ceiling only.*

Divide the length of the longest unsupported part in feet by the cube root of the breadth in inches, and the quotient multiplied by 1·47 will give the depth in inches.

3. *To find the dimensions of the king-post.*

Multiply the length of the post in feet by the span in feet: multiply the product by 0·12, which will give the area of the section of the post in inches. Divide this by the breadth for the thickness, or by the thickness for the breadth.

4. *To find the dimensions of the struts.*

Multiply the square root of the length of rafter supported in feet by the length of the strut in feet, and the square root of the product multiplied by 0·8 will give the depth, which multiplied by 0·6 will give the thickness.

The following table, copied from Hurst's revised edition of Tredgold's *Carpentry*, gives the scantlings for quarter-pitch king-post trusses of different spans, fixed 10 feet from centre to centre, the timber being "Baltic pine" (otherwise known as red or yellow deal or fir), and the roofs being covered with slates. The scantlings of the principal rafters are with one exception slightly larger than those found according to Tredgold's rules.

TABLE X.—SCANTLINGS FOR KING-POST ROOF-TRUSSES OF QUARTER-PITCH (ACCORDING TO TREDGOLD)

Span.	Tie-beam.	King-post.	Principal Rafters.	Struts.	Purlins.	Common Rafters.
Feet.	Inches.	Inches.	Inches.	Inches.	Inches.	Inches.
20	$9\frac{1}{2} \times 4$	4×3	4×4	$3\frac{1}{2} \times 2$	$8 \times 4\frac{3}{4}$	$3\frac{1}{2} \times 2$
22	$9\frac{1}{2} \times 5$	5×3	5×3	$3\frac{3}{4} \times 2\frac{1}{4}$	$8\frac{1}{4} \times 5$	$3\frac{3}{4} \times 2$
24	$10\frac{1}{2} \times 5$	$5 \times 3\frac{1}{2}$	$5 \times 3\frac{1}{2}$	$4 \times 2\frac{1}{2}$	$8\frac{1}{2} \times 5$	4×2
26	$11\frac{1}{2} \times 5$	5×4	$5 \times 4\frac{1}{4}$	$4\frac{1}{4} \times 2\frac{1}{2}$	$8\frac{3}{4} \times 5$	$4\frac{1}{4} \times 2$
28	$11\frac{1}{2} \times 6$	6×4	$6 \times 3\frac{1}{2}$	$4\frac{1}{2} \times 2\frac{3}{4}$	$8\frac{3}{4} \times 5\frac{1}{4}$	$4\frac{1}{2} \times 2$
30	$12\frac{1}{2} \times 6$	$6 \times 4\frac{1}{2}$	6×4	$4\frac{3}{4} \times 3$	$9 \times 5\frac{1}{2}$	$4\frac{3}{4} \times 2$

With a single exception, the compression members are oblong in section, and will offer less resistance than square timbers of the same sectional area; the departure from the square is not very pronounced in some of the principal rafters, but in the struts it is so great as to reduce the resistance very considerably. The sectional area of the tie-beam is on the average about three times that of the corresponding principal rafter, although the tensile stress in the tie-beam (due to the roof-load) is considerably less than the compressive stress in the rafter, and although timber has a greater resistance to tension than to compression; the extra size of the tie-beam is provided because this member is intended to carry a plastered ceiling, but we are not told what reduction in the scantling may be made if the ceiling is omitted. Similarly the king-post is subjected by the roof-load to a much less stress than the principal rafters, but Tredgold makes these members of approximately the same scantling; here again the weight of the ceiling is allowed for, and also the great reduction of the sectional area caused by making the joints.

Some experiments by Major G. K. Scott-Moncrieff, R.E., on full-size timber trusses, show clearly the unscientific nature of Tredgold's rules.[1] A truss 28 feet span was constructed with tie-beam $11\frac{1}{2} \times 6$ inches, king-post 6×4 inches, principal rafters $6 \times 3\frac{1}{2}$ inches, and struts $4\frac{1}{2} \times 2\frac{3}{4}$ inches. Loads were applied to the principal rafters immediately over the heads of the struts, that is to say, at the points where the purlins would be placed. "The truss showed signs of weakness at 6000 lbs. on each purlin. At this pressure the struts buckled and crushed the fibres at the foot of the king-post. When the pressure reached 7500 lbs. on each side, one of the principals [*i.e.* principal rafters] which had bent into a sinuous curve, split at a tight knot, and no further resistance was afforded. This truss showed great stiffness as a whole, on account of the heavy tie-beam. . . . The weakness of the truss lay in the insufficient scantling of the compression members, and in their unscientific proportions. . . . These struts buckled in the direction of their least dimension."

It will be of service to consider the result of this test a little more closely. The total weight under which it failed was 15,000 lbs. Tredgold's trusses are supposed to be strong enough, when placed at intervals of 10 feet, to carry a load of $66\frac{1}{2}$ lbs.[2] per square foot. For a span of 28 feet the length of each slope will be 16 feet. The total load on the roof will therefore be 2×16 feet $\times 10$ feet $\times 66\frac{1}{2}$ lbs. $= 21,280$ lbs. Deduct from this the proportion which would be transmitted directly to the walls by means of the common rafters —namely, $\frac{1}{4}$ of 21,280 lbs. $= 5320$ lbs.—and we have the net weight of 15,960 lbs., which the truss is supposed to be capable of bearing *with safety.* A smaller load than this (namely, 7500 lbs. on each side) proved to be the *ultimate breaking weight* of the truss, and some of the members were seriously crippled under a load of 12,000 lbs. It is true that the allowance of $66\frac{1}{2}$ lbs. per square foot errs on the safe side, and consequently in practice Tredgold's scantlings have been used without disaster, but the materials and workmanship must be of the best.

A modified form of king-post truss, as shown in fig. 739, was then tested by Major Scott-Moncrieff, and although some of the timbers were of lighter scantling, the truss did not actually fail under a load of 11,000 lbs. on each side, although it had begun to buckle under a load of 9000 lbs. on each side. It was again tested after "lying about exposed to wintry weather for some months", and failed under a total load of 19,100 lbs.; the reduction in strength was believed to be due to shrinkage at the foot of the rafters, which had rendered the joints there somewhat loose. Tredgold's truss had failed under a load of 15,000 lbs. The new truss had a span of 28 feet, with principal rafters 5×6 inches, struts 5×4 inches, king-post 5×2 inches at the smallest cross section, and tie-beam double, each flitch being 8×2 inches. The truss was designed so that the centres of pressure of the different members met at the joints, thus avoiding eccentric stresses. The joint at the foot of the king-post was secured by a strap, with gibs and cotters at the middle of the depth of the tie-beam. The joint at the end of the tie-beam was more novel. The principal rafter abutted against a hardwood block, which was rounded at the end and retained in position by a horizontal iron strap 2 inches wide, secured by a single 2-inch bolt passing through

[1] See the *Journal* of the Royal Institute of British Architects, January 14, 1899.

[2] We will assume that this includes the weight of the ceiling.

the flitches of the tie-beam and a hardwood block or distance-piece. It was afterwards found desirable to have some means of tightening the strap, and cotters were used instead of the bolt.

The stresses in a king-post truss are worked out graphically in Plate XXV. The trusses are supposed to be fixed 10 feet from centre to centre, and the pitch is one-fourth of the span. The dead load on the roof is calculated at 16 lbs. per square foot (fig. 1), and gives the stresses shown in fig. 2. If a ceiling weighing 10 lbs.[1] per square foot is added, the loads for roof covering and ceiling become as in fig. 3, and give the stress diagram fig. 4. The wind-pressure acts on one side only of the roof, and, if calculated at 30 lbs. per square foot, gives the forces shown in fig. 5 and the stress diagram fig. 6.

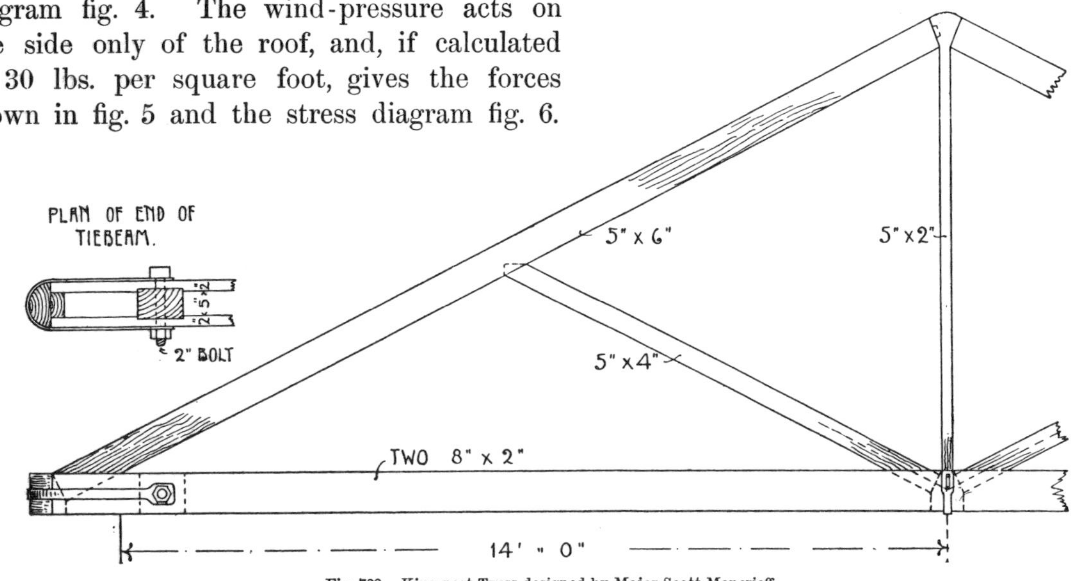

Fig. 739.—King-post Truss designed by Major Scott-Moncrieff

If the vertical components of the wind-pressure are taken as shown by the vertical dotted lines in fig. 5, we obtain the stress diagram fig. 7; this method is not generally considered as accurate as the other.

For the purposes of comparison with Tredgold's scantlings, we will consider a truss supporting a ceiling in addition to the roof covering. Wind-pressure must also be allowed for. This involves a combination of the stresses shown in figs. 4 and 6. Only that half of the truss which is most severely stressed need be considered, as the other half will be similarly stressed when the wind blows from the opposite quarter. Applying the scale to the stress diagrams, we obtain the following figures:—

Principal rafter B I,	+ 6200	+ 4400	=	+	10,600	lbs.
" " C K,	+ 4700	+ 2500	=	+	7,200	"
Strut I K,	+ 1500	+ 3130	=	+	4,630	"
King-post K L,	− 2840	− 1400	=	−	4,240	"
Tie-beam H I,	− 5560	− 4900	=	−	10,460	"

The sign + represents compression, and the sign − tension. We will assume that the breadth of the members is to be uniformly 4 inches. The total length of the principal rafter is about 200 inches, but as the rafter is prevented from lateral buckling by the purlins, it may be regarded as a post whose length, divided by the breadth, is $\frac{100 \text{ inches}}{4} = 25$. According to Plate XXVI, the ultimate resistance of a red-deal post of this ratio will be from 1840 to 2260 lbs. per square inch—say 2100 lbs. This resistance cannot, however, be regarded as probable for two reasons—(1) The ends of the rafters are not cut entirely square, and eccentric loading will therefore occur, and (2) the ends abut against wood.

[1] For ornamental ceilings a greater weight must be allowed for. It is also advisable to allow for the weight of workmen engaged in repairing the roof, and other loads which may occasionally be placed on the beam.

It cannot be stated accurately what deduction must be made in respect of these defects of construction, but we may assume that they will be covered by the use of a factor of safety of 7. The safe load, therefore, will be $\frac{2100}{7} = 300$ lbs. per square inch. The total load is 10,600 lbs., and the breadth of the rafter 4 inches. The depth must therefore be $\frac{10,600}{300 \times 4} = 9$ inches nearly.

The principal rafter is generally made of the same scantling from end to end, and the upper half need not be considered in detail, as it is stressed to a smaller degree than the lower.

The strut (allowing for the depth of the rafter and the breadth of the king-post) is about 90 inches long, and assuming for the moment that the smaller side of the cross section will be 4 inches, we find it to be $22\frac{1}{2}$ diameters in length. The ultimate resistance may therefore be read from Plate XXVI as about 2100 lbs. per square inch, and the safe resistance will be $\frac{2100}{7} = 300$. The total load is 4630 lbs., and the breadth 4 inches. The depth must therefore be $\frac{4630}{300 \times 4} = 4$ inches nearly.

The tensile stress in the king-post is 4240 lbs. If we consider the ultimate resistance of red deal to tension to be 8000 lbs. per square inch, and allow a factor of safety of 10, the area of the cross section will be $\frac{4240}{800} = 5\frac{3}{10}$ square inches; this will be given by a piece 4 inches by $1\frac{1}{3}$ inch. An addition must, however, be made for the hole cut through the king-post to receive the gibs and cotters—say 1 inch. This will increase the scantling to 4 inches by $2\frac{1}{3}$ inches. In practice, however, larger scantlings must be adopted, as the head and foot of the king-post must be of large size to furnish abutments for the principal rafters and struts, and as long pieces free from knots cannot be obtained. The ultimate shearing resistance of red deal is about 600 lbs. per square inch, and the safe stress may be taken to be 120 lbs. The total shearing areas at each end of the king-post must therefore be $\frac{4240}{120} = 35\frac{1}{3}$ square inches. There will be two planes of shearing in each case, and each plane must contain $17\frac{2}{3}$ square inches. As the breadth is 4 inches, the length must be $\frac{17\frac{2}{3}}{4} = 4\frac{1}{2}$ inches nearly. This gives us the distance from the tie-beam to the shoulders formed in the king-post to receive the struts, and also the length of the enlarged head of the king-post. Shoulders of sufficient size to receive the ends of the struts and principal rafters can be obtained by making the king-post 7 inches by 4 inches.

Proceeding in a similar manner, the sectional area of the tie-beam required to resist the tensile stress of 10,460 lbs. will be $\frac{10460}{800} = 13\frac{3}{40}$ sq. inches, which will be furnished by a scantling of 4 inches by $3\frac{1}{3}$ nearly. The tie-beam is, however, cut to receive the ends of the king-post and principal rafters, and must be increased in size accordingly. If we allow 3 inches for the depth of the notch to receive each rafter, we shall require a piece 4 inches by $6\frac{1}{3}$ inches.

But in this case, the tie-beam is also loaded transversely by the ceiling. Each half is a beam, which may safely be regarded as fixed at one end and supported at the other, and loaded with 1500 lbs., equally distributed. The scantling can be calculated from the formula

$$W = \frac{4 f b d^2}{3 l}, \text{ or } d^2 = \frac{3 W l}{4 f b}.$$

Taking the value of f at 4200 lbs., and allowing a factor of safety of 6, we get the safe value of $f = 700$ lbs. Then

$$d^2 = \frac{3 \times 1500 \text{ lbs.} \times 180 \text{ inches}}{4 \times 700 \times 4} = 72{\cdot}32, \text{ and}$$

$$d = \sqrt{72{\cdot}32} = 8\tfrac{1}{2} \text{ inches.}$$

A beam $8\frac{1}{2}$ inches by 4 inches is therefore required to carry the ceiling only. The net scantling required to resist the tensile stress was found to be 4 inches by $3\frac{1}{3}$ inches. Add these together and a scantling of $8\frac{1}{2}$ inches by $5\frac{1}{2}$ inches is obtained. Instead of increasing the breadth only, the depth also may be increased, so that the scantling becomes 10 inches by $4\frac{3}{4}$ inches (say 10 inches × 5 inches); the deeper beam has the advantage of greater strength and stiffness.

The dimensions of the various members will therefore be:—Principal rafters, 9 inches by 4 inches; king-post, 7 inches by 4 inches; tie-beam, 10 inches by 5 inches; and struts, 4 inches by 4 inches. If the principal rafters had been of square section, a smaller quantity of timber would have been required, but this would probably not have reduced the cost, as thicker scantlings of Baltic wood fetch higher prices per cubic foot or per standard. It will be seen that the principal rafters and struts are of greater dimensions than those recommended by Tredgold for the same span, although a smaller load is allowed for.

In mediæval examples, the king-post was almost invariably designed as a compression member, transmitting the load on the ridge directly to the tie-beam, and the latter was made of sufficient strength to carry the roof-load, as if it were a simple beam. This accounts for the application of the terms "post" and "beam" to these members. An extreme example is the chancel roof of St. Martin's Church, Leicester, illustrated in Brandon's *Open Timber Roofs of the Middle Ages*. It has the appearance of a truss, but there are no principal rafters, and the construction is simply a combination of beams and posts. The tie-beam, which is stiffened by curved braces, is supported by wall-posts, and is really a beam by which the king-post and two short queen-posts or struts are carried; these in turn support the ridge-piece and purlins, on which the common rafters rest. The span of the roof is 23 feet, and the "trusses" are 12 feet 6 inches apart. The tie-beam is 20 × 14 inches, purlins 10 × 10 inches, ridge-piece 10 × 9 inches, cornice 10 × 8 inches, and common rafters $6\frac{1}{2}$ × $4\frac{1}{2}$ inches. The cornice is framed into the tie-beams and does not rest upon the wall; it is stiffened by arched braces rising from the wall-posts. A wall-plate receives the lower ends of the common rafters, the upper ends being framed into the ridge.

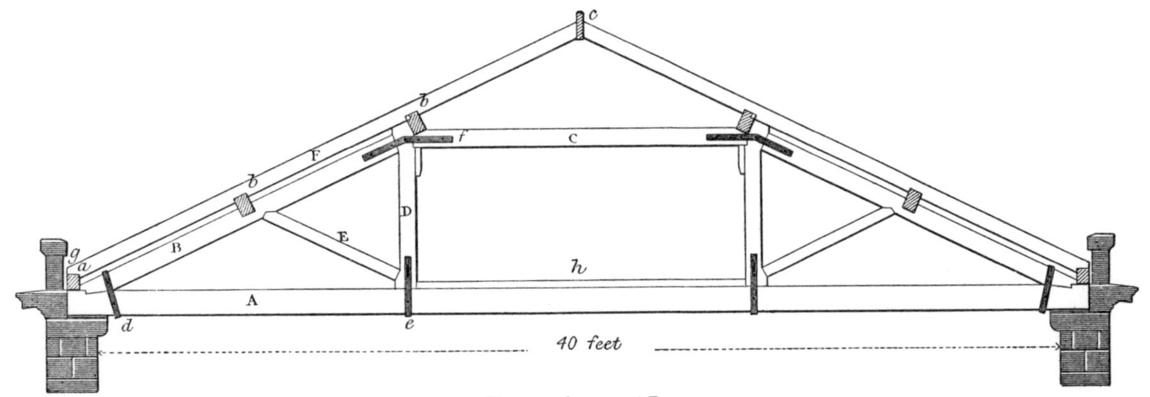

Fig. 740.—Queen-post Truss

II. *Queen-post Trusses.*—Trusses of this kind are usually considered better than king-post trusses for spans exceeding 30 feet. We have already seen (pages 328–9, Vol. I) that the queen-post truss is not well adapted for unequal loads, such as that caused by the wind acting on one side of the roof; in such a case the tie-beam is subjected to a transverse stress in addition to the tensile stress, and if the principal rafters are continued to the apex, they also are transversely stressed. The defect can be avoided, as already explained, by one or two diagonal members fixed across the central rectangle of the truss, but such members have the disadvantage of interfering with the open space in the middle of the span.

An example of ordinary type for a span of 40 feet is given in fig. 740. The scantlings are as follows:—A, tie-beam, 12 × 6 inches; B, principal rafter, 10 × 6 inches; C, straining-beam, 9 × 6 inches; D, queen-post, 8 × 6 inches; E, strut, 6 × 6 inches; F, common rafter, 6 × $2\frac{1}{2}$ inches. Nearly all the members are stronger than is actually necessary. The pole-plate is shown at *a*, the purlins at *b b*, and the ridge-piece at *c*; the strap *d* at the foot of the principal rafter is too nearly vertical to be of much use; *e* is the strap at the foot of a queen-post, and is provided with gibs and cotters; *f* is the strap at the head of the queen-post, secured by bolts to the three members forming the joint; *g* the parapet gutter, and *h* the straining sill. Cleats are nailed to the queen-posts to assist in supporting the straining-beam C. The tie-beam is supported in two places by the queen-posts, and two purlins can be inserted without bringing a transverse stress on the principal rafters. In a truss framed in this manner, about one-half of the dead load on each slope is concentrated at the joint at the head of the queen-post, about one-third at the head of the strut, and the remainder at the foot of the truss. Unless special circumstances have to be considered, the position of the purlins should first be fixed so as to divide the common rafters into three equal spans; the truss can then be drawn in such a manner that the purlins lie over the heads of the queen-posts and struts.

Tredgold's rules for finding the scantlings of queen-post trusses are, like his rules for king-post trusses, of an empirical nature; they are as follows for trusses, the tie-beams of which support plastered ceilings:—

1. *To find the dimensions of the principal rafters.*

Multiply the square of the length of the rafter in feet by the total span in feet, and divide the product by the cube of the thickness in inches: the quotient multiplied by 0·155 will give the depth.

2. *To find the dimensions of the tie-beam.*

Divide the length of the longest unsupported part (in feet) by the cube root of the breadth (in inches), and the quotient multiplied by 1·47 will give the depth.

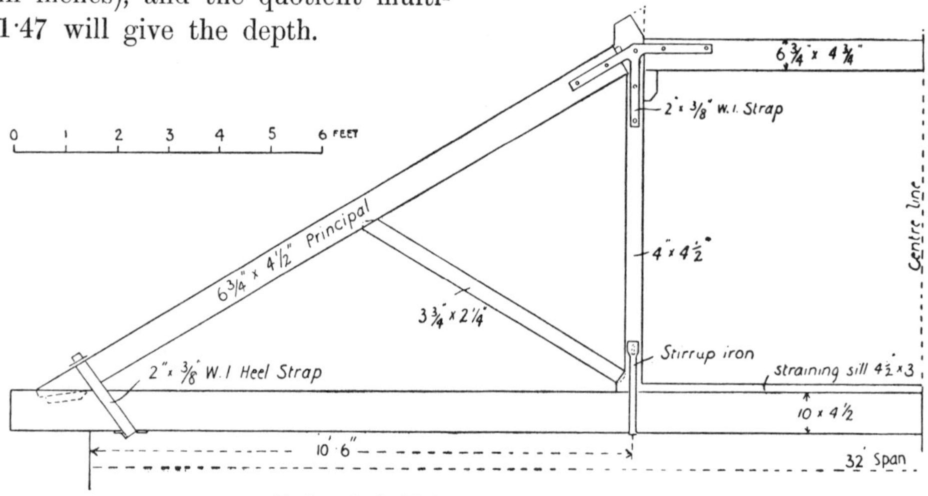

Fig. 741.—Tredgold's Queen-post Truss, 32 ft. span

3. *To find the dimensions of the queen-posts.*

Multiply the length in feet of the post by the length in feet of that part of the tie-beam which it supports: the product multiplied by 0·27 will give the area of the post in inches; and the breadth and thickness can be found as in the king-post.

4. The dimensions of the struts are found as for king-post trusses.

5. *To find the dimensions of the straining-beam.*

Multiply the square root of the span in feet by the length of the straining-beam in feet, and extract the square root of the product: multiply the result by 0·9, which will give the depth in inches. To find the breadth, multiply the depth by 0·7.

Major Scott-Moncrieff's tests of full-size timber trusses included two queen-post trusses, 32 feet span, the first (fig. 741) constructed according to Tredgold's rules, and the second (fig. 742) with the tie-beam composed of two flitches secured to the rafters and queen-posts with straps and cotters. The scantlings and details are given in the illustrations. Unfortunately the trusses buckled laterally to such an extent that the tests were far from satisfactory. Of the two Tredgold's appears to have been the stiffer laterally, due probably to the greater amount of iron at the joints, but the cost of this truss was £4, 10*s.* 3*d.*, against

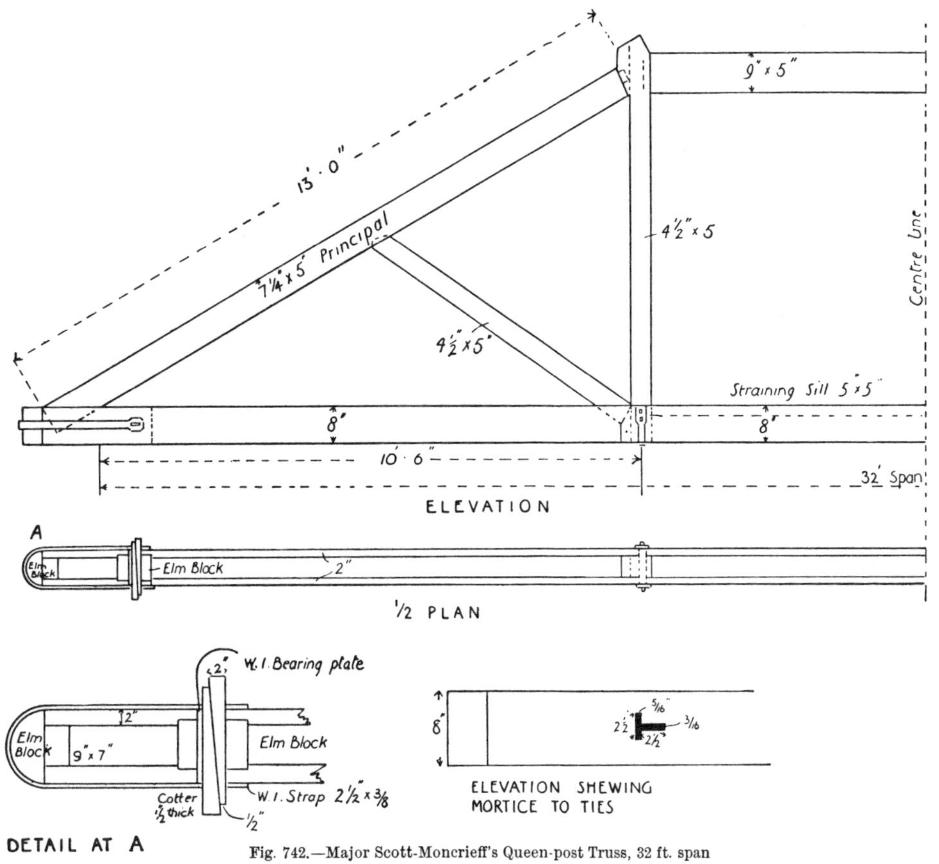

Fig. 742.—Major Scott-Moncrieff's Queen-post Truss, 32 ft. span

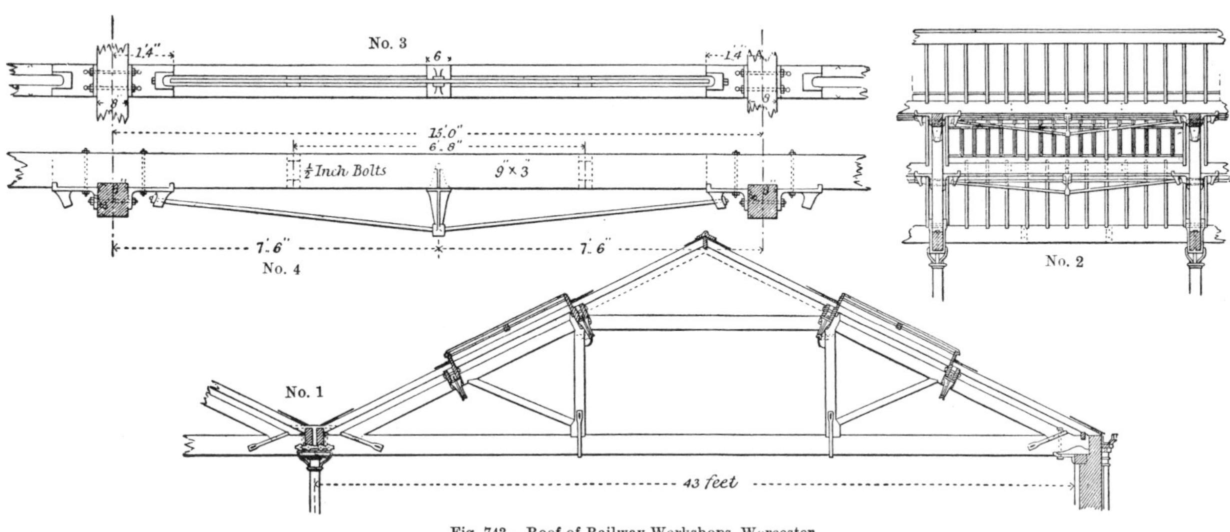

Fig. 743.—Roof of Railway Workshops, Worcester

£3, 14*s.* 8*d.* for Major Scott-Moncrieff's truss. At a total pressure of 24,780 lbs. Tredgold's truss showed signs of fibre-crushing at the ends of the struts, and of shearing along the grain in the queen-posts below the struts, and at 31,920 lbs. the joint at the foot of the rafter was crippled, the fibres being crushed and the strap strained and bent. This is equivalent to a load of about 100 lbs. per square foot for trusses fixed 10 feet apart, or only 50 per cent more than the safe load for which Tredgold designed his trusses. This is obviously a very low factor of safety. Major Scott-Moncrieff's truss gave promise of greater strength, and no failure whatever was observed at the joints under a load of 30,000 lbs., but the lateral twisting was so great as to overturn the temporary props, and the truss was damaged in falling.

Nos. 1, 2, 3, and 4, fig. 743, are the elevation and details of the queen-post roof of the railway workshops at Worcester. The principals were placed 15 feet apart, and the purlins

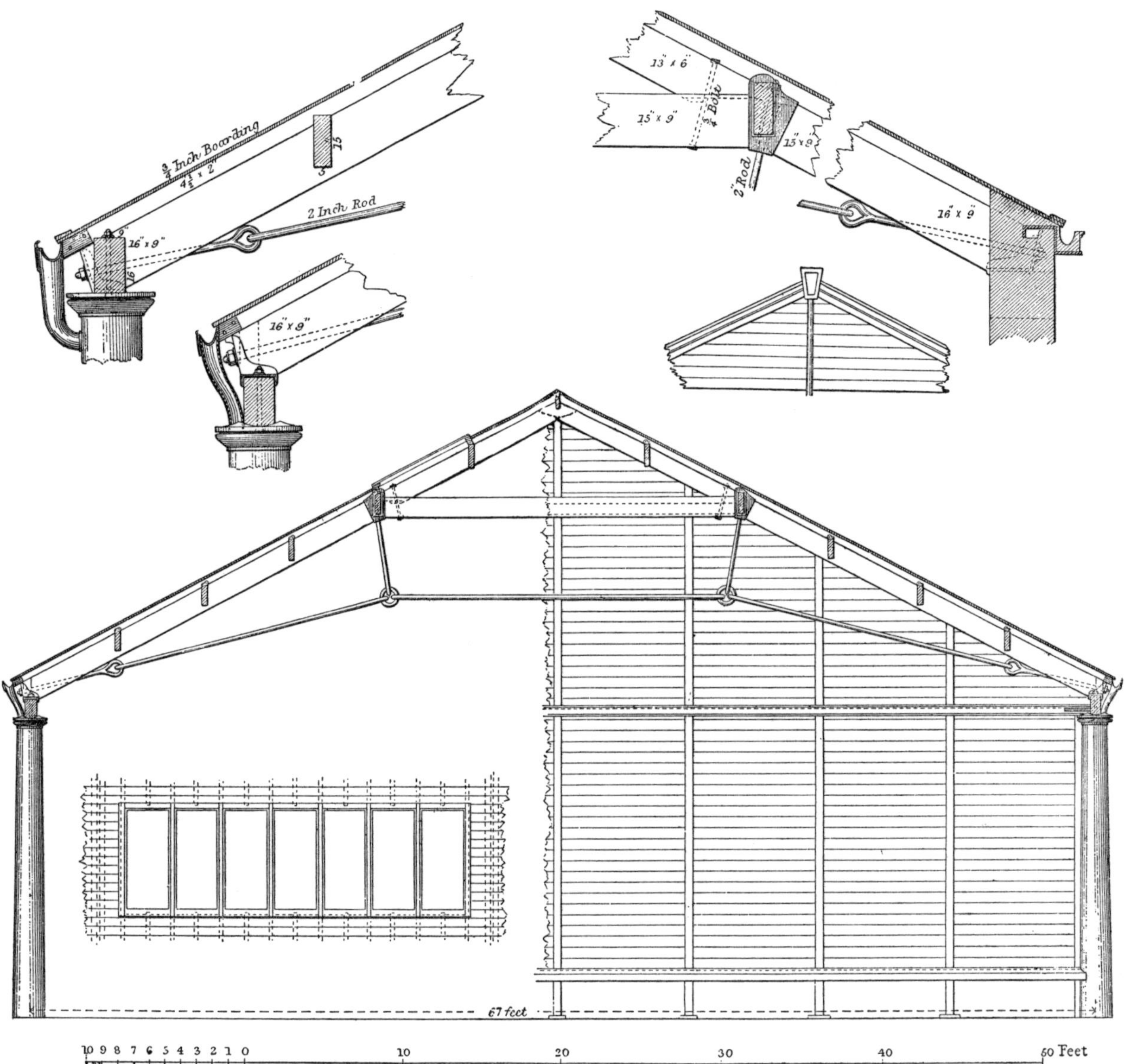

Fig. 744.—Roof, Salthouse Dock, Liverpool

were trussed. The details are as follows:—Principal rafters, 8 × 8 inches; tie-beam, 12 × 8 inches; queen-post, 8 × 6 inches; struts, 4½ × 4½ inches; straining-beam, 9 × 8 inches; common rafters, 4½ × 2 inches; purlins, in two flitches, each 9 × 3 inches. The tie-beams are carried on iron shoes. No. 1 is the elevation of the truss. No. 2 is a section showing one bay of the roof. No. 3 shows the soffit, and No. 4 the side of one of the trussed purlins drawn to a larger scale.

A queen-post truss with wrought-iron tie-rods and queen-rods is illustrated in fig. 744. It was designed by Jesse Hartley, and erected over the shed on the east quay of the Salthouse Dock, Liverpool. The principal rafters are shown in two lengths jointed at the head of the queen-rods, the lower parts 16 × 9 inches, and the upper parts 13 × 6 inches; the straining or collar beam is 15 × 9 inches, and the rods are all 2 inches in diameter. The purlins are 15 × 5 inches, and the common rafters $4\frac{1}{2}$ × 2 inches. The trusses are not well designed. The principal rafters are transversely stressed to a serious extent, but the principal defect is the absence of any provision for unequal loading. We have seen that an ordinary queen-post is prevented from distortion, under unequal loading, by the stiffness of the tie-beam, or, if the rafters are continued to the ridge, by the stiffness of the tie-beam and rafters. In this case, a tie-rod, entirely lacking in transverse strength, is substituted for the tie-beam, and the rafters are jointed at the heads of the queen-rods, so that no strength is provided at these points. These defects appear to have rendered the trusses unstable. The writer visited the shed in 1901, and found that additional tie-rods had been inserted from the centre of the collar-beam to the links connecting the main tie-rods and queen-rods. These new rods are about $1\frac{1}{2}$ inch in diameter, and are unsatisfactory in so far as they cause a transverse stress in the collar-beam. A brick wall has also been built along one side of the shed, and other alterations made.

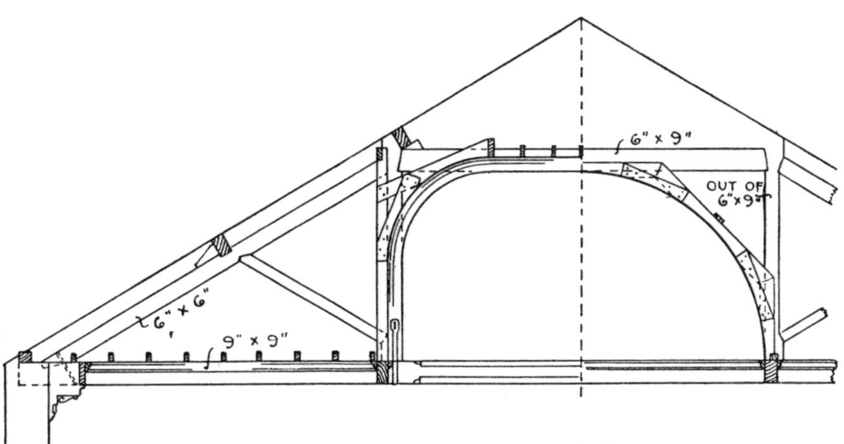

Fig. 745.—Coved Ceilings in Central Portion of Queen-post Roof, 36 ft. span

Fig. 745 shows two kinds of coved ceiling formed in the central portion of a queen-post truss. This is a convenient way of increasing the air-space and height of a somewhat low room, and has been adopted by the writer in several chapels. The wood-work of the truss may be exposed to view or may be blocked out as required and covered with laths and plaster, or with fibrous plaster. In the example given the wood framework is exposed. Braces, following the curves of the coves, are introduced between the queen-posts and straining-beams, and longitudinal bearers are framed between the tie-beams at the feet of the queen-posts. On these bearers the studs and coves are supported. Another method of utilizing the central space is shown in fig. 746. In this case the cove is formed between the queen-post and the wall, and the central space is arched. The outer portions of the truss may be concealed by plain or panelled boarding, or the struts may be exposed as in the example, where an additional strut is inserted for the sake of effect. As the plastered coves increase the load on the truss, the scantlings of the members ought to be somewhat larger than usual.

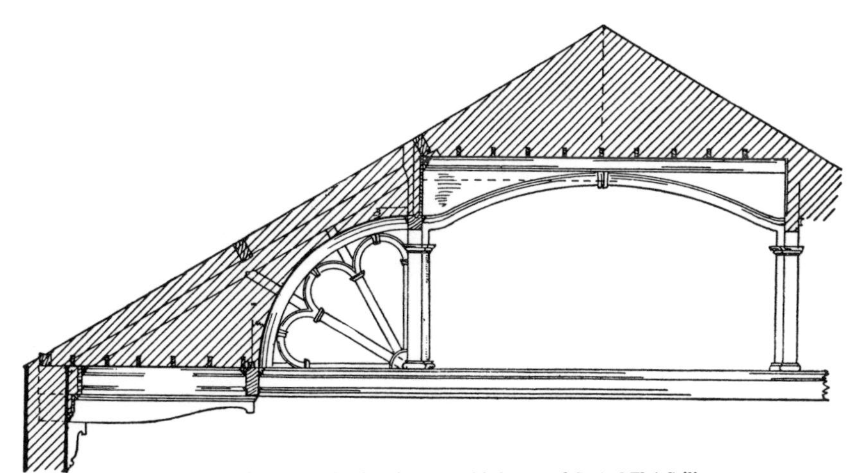
Fig. 746.—Queen-post Roof, 36 ft. span, with Coves and Central Flat Ceiling

The ordinary queen-post truss is divided into three approximately equal parts by the queen-posts, and the central coves, as shown in fig. 745, are too narrow in proportion to the width of the room. It is better to divide the span into four parts, giving two to the central space and one to each side space. The principal rafters in this case will be continued to the ridge, and the queen-posts and straining-beam may be composed of two flitches, one on each side of the rafters. The part of the truss above the straining-beam can be framed as a king-post truss, the general arrangement of the members being like fig. 3, Plate XXIX, but without the post marked E between the straining-beam and tie-beam.

III. *Compound King-post and Queen-post Trusses.*—In order to avoid transverse stress on the principal rafters it is desirable to introduce additional struts in trusses of large span. This necessitates additional suspending posts or rods. A truss containing a king-post and two queen-posts, as shown in fig. 3, Plate XXVII, is a common type. The scantlings of this example as executed, and as determined by Tredgold's rules, are given below for purposes of comparison:—

	As Executed.	As Determined by the Rules.		As Executed.	As Determined by the Rules.
Tie-beam	11 × 9 ins.	$11\frac{1}{2}$ × 6 ins.	Principal struts	6 × 6 ins.	6 × 3 ins.
King-post	10 × 8 „	8 × 6 „	Secondary struts	5 × 5 „	3 × 2 „
Suspending posts or queen-posts ...	10 × 7 „	5 × 3 „	Purlins	7 × 9 „	
			Cleats at back of purlins	8 × 6 „	
Principal rafters ...	10 × 7 „	8 × 6 „	Ridge-piece	8 × $1\frac{1}{2}$ „	

On comparing the scantlings of this roof as executed, with those derived from the application of Tredgold's rules, there will be found an excess of strength. The scantlings, in point of fact, are nearly as large as those of other examples, of much greater span, which are here given. The tie-beam is strapped to the king and queen posts, and the principal rafters are secured by screwed bolts and nuts.

This method of trussing lends itself easily to a combination of wood struts and wrought-iron or steel vertical tie-rods. Fig. 747 contains examples carried out from the writer's designs for the roofs of dye-works. The trusses Nos. 1 and 2 are over different parts of the same room. The eaves and ridge are level throughout, and the ridge is in the same straight line. The trusses over the narrower part of the room have therefore one side of the normal pitch and the other much quicker. The half truss shown in No. 3 has the tie-beam and principal rafter supported by walls at both ends. The trusses are 13 feet between the centres, and together with the purlins are entirely of pitch pine. As it was considered probable that the tie-beams would be used to support loads in addition to the normal roof-load, they were made stronger than would otherwise have been necessary.

A roof of similar character, but exhibiting a freer use of cast-iron sockets, is shown in fig. 1, Plate XXVII. The head of the king-rod passes through a cast-iron socket at A, shown to a larger scale in No. 2; the lower end passes through the tie-beam and is secured with a nut. The lower ends of the queen-rods pass through castings which serve as abutments for the struts. To avoid cutting the principal rafters cast-iron sockets or stirrups are bolted to them to receive the ends of the purlins. Cast-iron shoes are laid on the walls to receive the ends of the tie-beam and principal rafters, the sole-plate being prolonged to admit of its being bolted to the tie-beam. Fig. 2 shows the same method of trussing developed to a further degree. In this case purlins are dispensed with, the common rafters being laid horizontally and supported directly by the principal rafters. To avoid excessive transverse stress on the latter members, the number of struts and suspending rods is increased. The tie-beams, principals, and struts are first framed together; the suspending rods are then introduced, and screwed up by the nuts at their lower end until the framing is firmly united. A roof of this construction, 54 feet span and 212 feet long, was erected many years ago over the passengers' shed of the Croydon railway-station.

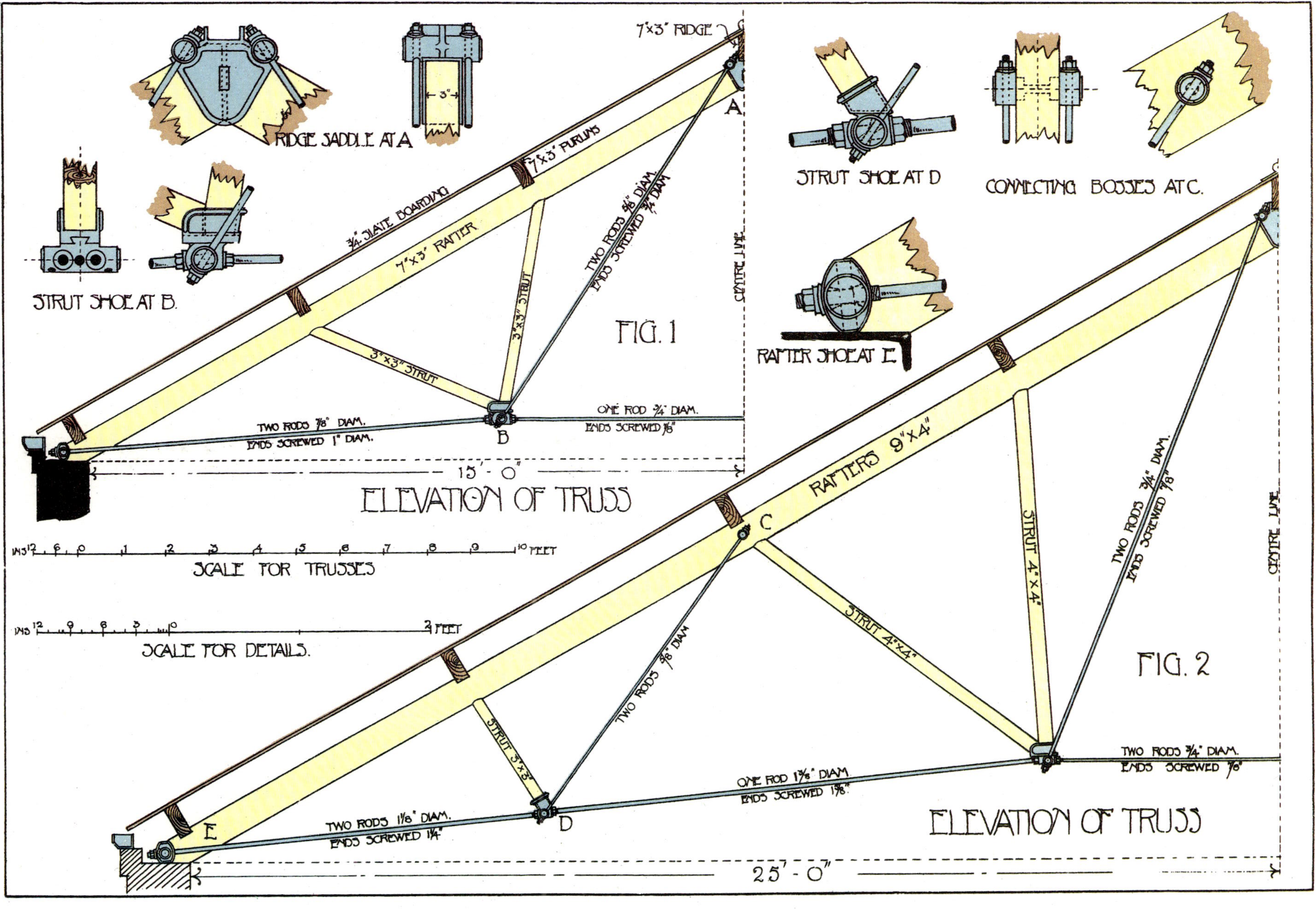

Mr. H. P. Holt's Wood-and-Iron Roof-Trusses

Plate XXVII, fig. 4.—No. 1 is the elevation of one of the trusses of George Heriot's School, Broughton Street, Edinburgh, designed by Alexander Black, architect, and executed under his superintendence. It is a clever example of trussing, and was in good conditior in 1901, more than forty years (I believe) after its erection.

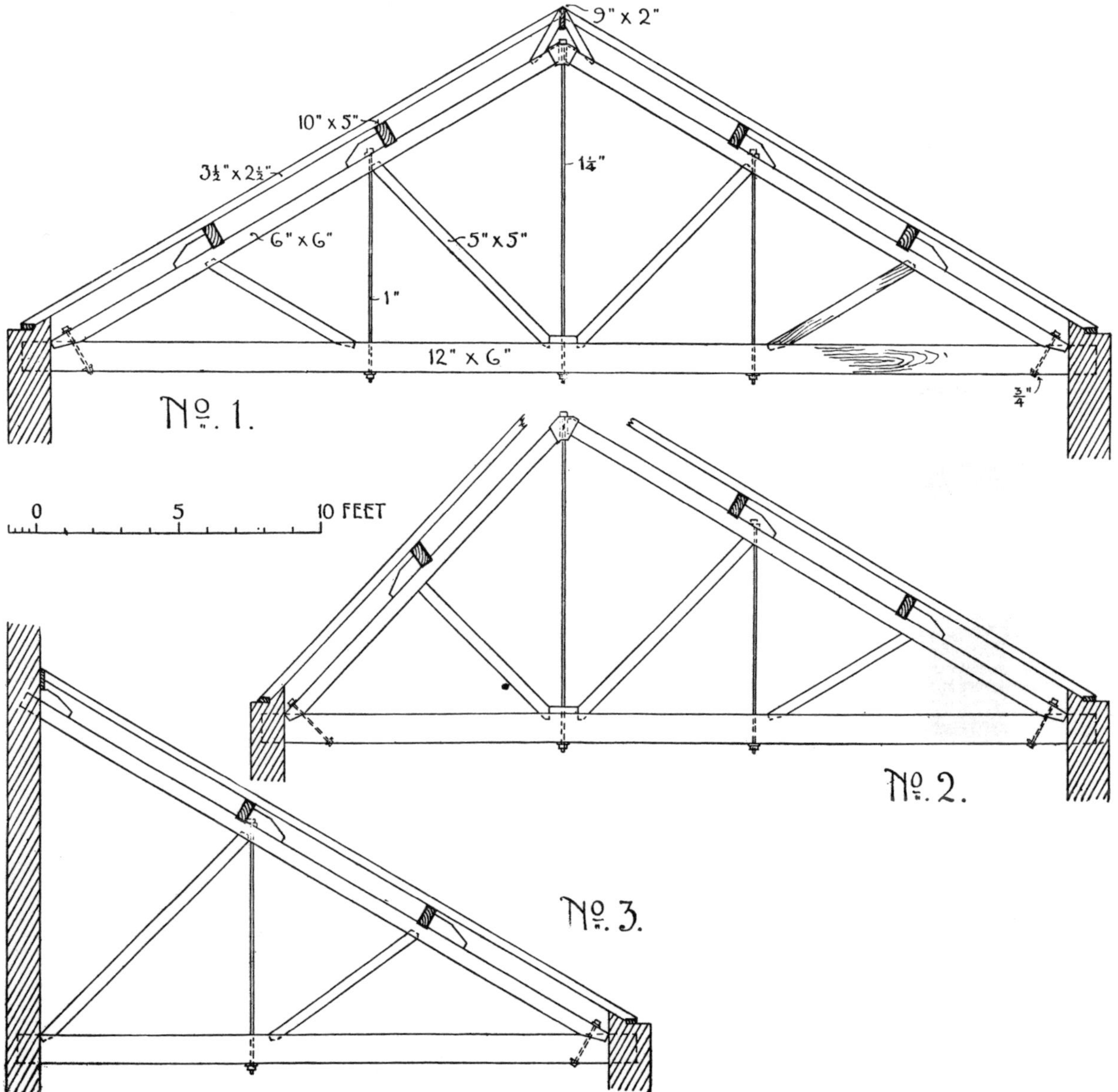

Fig. 747.—Two Compound King-post Trusses and Half Truss

The dimensions of the scantlings by Tredgold's rules are here placed in juxtaposition with the dimensions of the scantlings as executed, as follows:—

	By the Rule.	As Executed.		By the Rule.	As Executed.
A, Tie-beam	$11\frac{3}{4}$ × 4 ins.	12 × 4 ins.	E, King-post (oak) ...	8 × 4 ins.	5 × 4 ins.
B, Principal rafters ...	8 × 4 „	8 × 4 „	F, Queen-posts (oak)	5 × 5 „	4 × 4 „
C, Common rafters ...	...	5 × $2\frac{1}{2}$ „	G, H, Inner struts ...	$4\frac{3}{4}$ × $3\frac{1}{2}$ „	4 × 4 „
D, Purlins	9 × $5\frac{1}{2}$ „	9 × 6 „	I, Outer struts ...	3 × $2\frac{1}{2}$ „	4 × $3\frac{1}{2}$ „

No. 2 shows the details of the king-post to a larger scale. The heads of the rafters and the feet of the struts are received by cast-iron sockets bolted to the king-post. The tie-beam is suspended to the post by an iron strap with wedges. Cleats marked K, out of 6 × 4-inch stuff, are spiked to the principal rafters in order to keep the purlins in position. A section through the tie-beam, king-post, and strap is given at No. 3.

Nos. 4 and 5 show the scarfing of the purlins; and Nos. 6 and 7 the end of the tie-beam, with the iron shoe which receives the foot of the rafter, and the strap which secures it.

The truss shown in fig. 1, Plate XXVIII, contains two queen-posts and two shorter "posts", known as princess-posts. The span is 60 feet, and the scantlings are:—

Principal rafters	11 × 6 ins.	Princess-posts A	$3\frac{1}{2}$ × 3 ins.
Tie-beam	$12\frac{1}{4}$ × 6 „	Struts (large)	$4\frac{1}{2}$ × 3 „
Queen-posts B	8 × 6 „	Struts (small)	$3\frac{1}{2}$ × $2\frac{1}{2}$ „.

Additional members are inserted in the next example (fig. 2) to support the platform roof and the central part of the tie-beam, and have the advantage of rendering the truss suitable for unsymmetrical loads. The tie-beam is scarfed at *a* and *b*, and also strengthened by a fish-plate. The purlins *e*, secured to the heads of the queen-posts, and the piece *d*, carry the platform joists A, which are cambered to give the necessary slope to the lead covering.

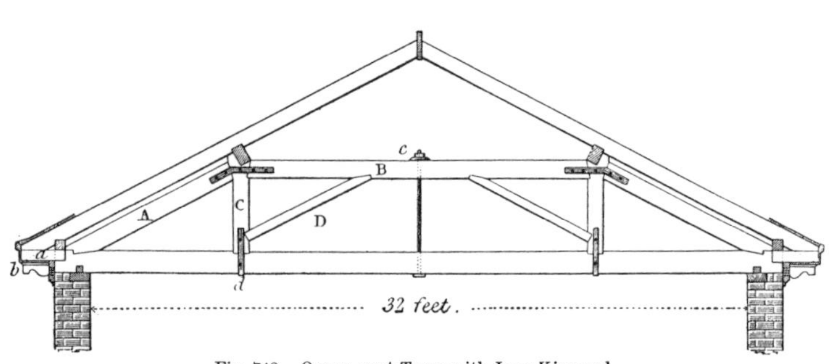

Fig. 748.—Queen-post Truss with Iron King-rod

Fig. 748 represents a queen-post roof, with an iron king-rod, intended for a span of 32 feet.

A, Principal rafter	11 × 5 inches.	C, Queen-post	9 × 5 inches.
B, Straining-piece	11 × 5 „	D, Struts	5 × 4 „

The common rafters are 8 × 3 inches, and project over the walls to form a cornice: *a* is the short ceiling joist of the cornice; *b*, an ornamental bracket; *c*, the king-bolt, and *d*, one of the straps.

A truss of somewhat similar design, for a span of 47 feet, is given in fig. 3, Plate XXVIII, but the common rafters are arranged in two ridges with a central gutter between, the latter being carried on two plates *c*, notched on to the straining-beams. The dimensions of the timbers, &c., are:—

A, Tie-beam	13 × 6 ins.	G, Common rafter	6 × $2\frac{1}{2}$ ins.
B, Principal rafter	11 × 6 „	*a*, Wall-plate for common rafters ...	7 × 9 „
C, Straining-beam	11 × 6 „	*b b*, Purlins	11 × 6 „
D, Queen-post	7 × 6 „	*e g*, King-bolt	2 ins. diameter.
E, Strut	7 × 6 „	*d*, Ridge-rafter	12 × $2\frac{1}{2}$ ins.
F, Counter-brace	6 × 6 „	*h*, Gutter-bearer	3 × $2\frac{1}{2}$ „

In this roof the purlins are shown framed into the principals, a practice which has already been censured.

The truss shown in fig. 4 has details similar in the main to those in fig. 1, Plate XXVII, but the arrangement of the members is different. The head of the principal rafter, and the end of the straining-beam, are inserted into a cast-iron socket, an elevation of which is seen, enlarged, at No. 1. The suspension rod A D passes through the solid part of the socket. It has a head at its upper end, and at its lower end it is screwed, and secured by a nut. On the side of the socket is cast a rest for the end of the purlin *a b*. To avoid cutting the principal rafters, the other purlin at B is also carried in a cast-iron rest bolted to the rafter. The centre suspending rod at E passes through a cast-iron socket, which serves as an abutment to the two main struts. Similar abutments are provided for the lower ends of the struts.

Fig. 1.

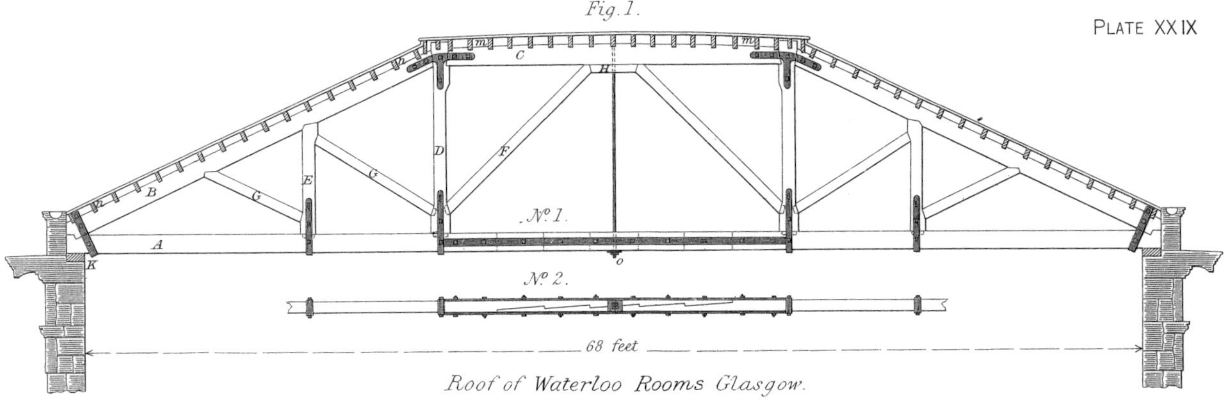

Roof of Waterloo Rooms Glasgow.

Fig. 2.

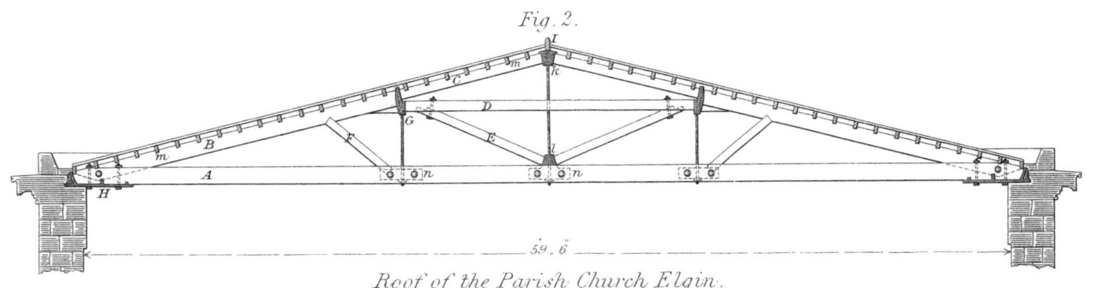

Roof of the Parish Church Elgin.

Fig. 3.

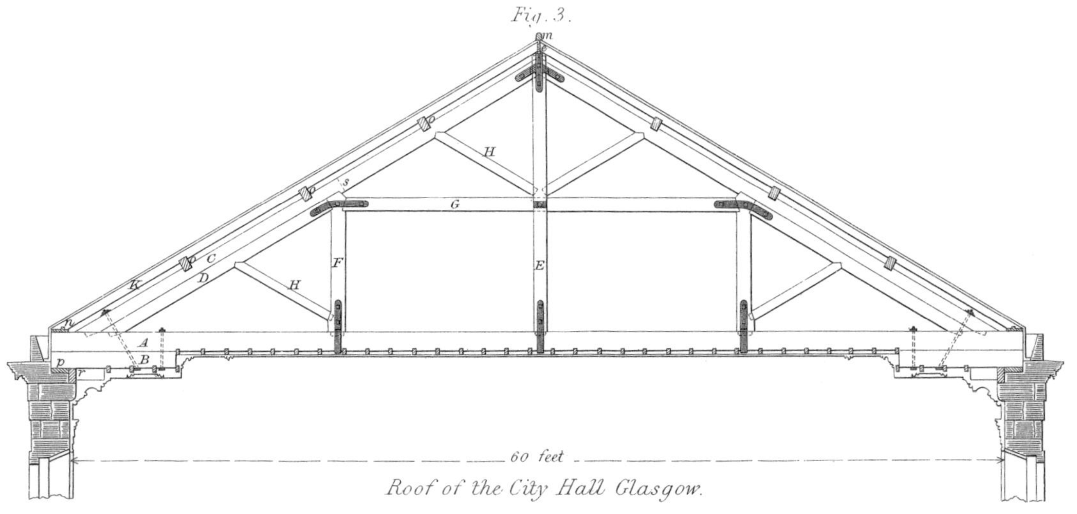

Roof of the City Hall Glasgow.

Fig. 4.

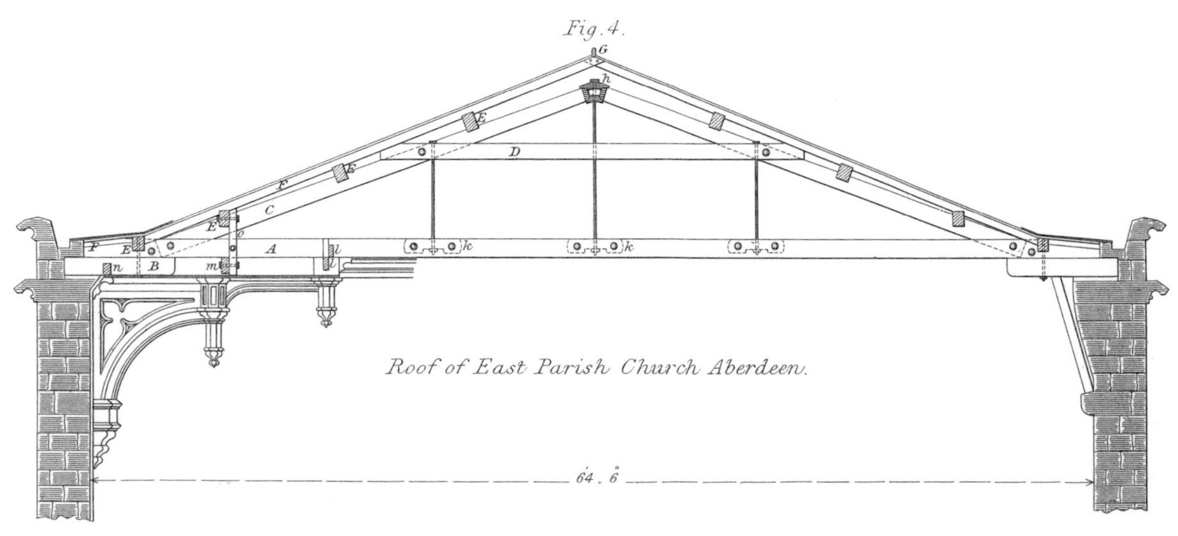

Roof of East Parish Church Aberdeen.

10 9 8 7 6 5 4 3 2 1 0 10 20 30 40

COMPOUND ROOF TRUSSES

Plate XXIX, fig. 1.—This represents one of the roof-principals of the Waterloo Rooms (formerly the Wellington Street United Presbyterian Church), Glasgow, which was designed by John Baird, architect, and erected in 1823; No. 1, elevation of one principal; No. 2, part of the upper side of the tie-beam:—

A, Tie-beam, 12 × 9 inches.
B, Principal rafter, 13 inches at bottom, 11 inches at top, and 9 inches thick.
C, Straining-beam, forming support of platform, and cambered, 13 inches deep at centre, 11 inches at ends, and 9 inches thick.
D, Principal queen-posts, 13 × 9 inches at top and bottom, and 9 × 9 in smallest part.
E, Second queen-posts (sometimes known as princess-posts), 10 × 9 inches at top and bottom, and 7 × 9 in smallest part.
F, Principal strut, 9 × 9 inches.
G G, Secondary struts, 7 × 9 inches.
H, Straining-piece between principal struts, 6 × 9 inches.
m m, Platform joists, 10 × 2½ inches, and 15 inches from centre to centre, covered with boarding 1¼ inch thick for lead.
n n, Common rafters, laid horizontally, 6 × 2½ inches, covered with slate-boarding ⅞ inch thick, and slates.

The ironwork is of the following dimensions:—

Straps at feet of principal rafters, 2½ × ⅝ inches.
Three-tailed straps connecting principal rafters, queen-posts, and straining-beam, 2½ × ⅝ inches; and each tail 2 feet 6 inches long.
Straps at feet of queen-posts, 2½ × ½ inches, bolted and keyed.
King-bolt 1¼ inch diameter, screwed up hard.

The principals are placed 9 feet 4 inches apart, each of the two outer ones forming a support for two half-principals, which are framed into it to carry the rafters along the hip. All the timbers are joined by mortise-and-tenon joints. The platform joists and horizontal rafters are notched on the straining-beams and principals. The tie-beams are in two lengths of timber, scarfed, as shown in No. 2. The scarf is secured by iron straps, each 3 inches wide and ⅝ inch thick, and bolted. A deeply-coffered plaster ceiling is carried by the principals. The writer inspected the roof in 1901, and to all appearance it was as sound as when it was erected 78 years before.

Fig. 2.—Elevation of a roof-principal of the parish church, Elgin, by Simpson of Aberdeen. This is a somewhat curious example of trussing for a low-pitched roof.

There are twelve principals similar to the elevation in fig. 2. They are placed 6 feet 6 inches apart between centres; and the scantlings are as follows:—

A, Tie-beam, in two flitches, each 13 × 5½ inches.
B C, Principal rafters, 11 inches deep at lower end, 8 inches at top, and 6 inches thick.
D, Collar-beam, 7 × 5½ inches.
E, Struts, 5 × 5½ inches.
F, Struts, 5 × 4¼ inches.
m m, Horizontal rafters, 4½ × 2½ inches, 13 inches apart, and covered with grooved-and-tongued deal 1 inch thick, and lead weighing 7 lbs. to the superficial foot.

The tie-beams have cast-iron shoes H at each end, with abutments formed for the rafters, and secured with ⅞-inch diameter bolts, with nuts and washers. The suspending rods are 1 inch square, and have abutment pieces for the rafters and struts.

Fig. 3.—One of the roof-principals of the City Hall, Glasgow, erected about 1840. The following are the dimensions of its timbers:—

A, Tie-beam, 14 × 12 inches.
B, Sill piece, 12 × 12 inches.
C, Principal rafter, at the end 8 × 7 inches, and at S 11 inches deep.
D, Ditto, where doubled at lower end, 8 × 7 inches.
E, King-post, in two thicknesses, each 10 × 6 inches.

F, Queen-posts, at top 13 × 7 inches, in middle 10 × 7 inches, and at bottom 13 × 12 inches; a piece $2\frac{1}{2}$ inches thick has been sawn from each side of the queen-post, above the horizontal line shown in the illustration a little above the strap.
G, Straining-beam, 10 × 7 inches.
H, Struts, 6 × 6 inches.
K, Common rafters, 6 × $2\frac{1}{2}$ inches, covered with boarding.
m, Ridge board, 10 × 2 inches; batten over it, rounded for lead, $3\frac{1}{2}$ × 3 inches.
n, Wall-plates, under common rafters, 12 × $1\frac{1}{2}$ inches, with firring strip, 2 × 2 inches.
o o, Purlins, 8 × 5 inches.
p, Outer wall-plates, 14 × 3 inches.
r, Inner wall-plates, resting on corbels, 11 × 5 inches.
Ceiling joists, 3 × 2 inches, resting on top of tie-beams; those shown represent the cradling under the beams for plaster.

The iron straps are 4 inches broad by $\frac{3}{8}$ inch thick, with bolts $\frac{3}{4}$ inch square. The bolts securing the ends of the rafters and the beams are 1 inch square, and their washers are the full breadth of the beams.

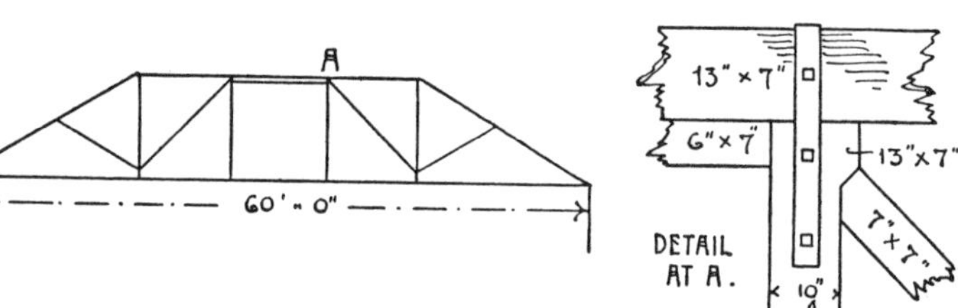

Fig. 749.—Outline Diagram of Truss for Hip of the City Hall, Glasgow, and Detail of Joint at A

The principals are placed 12 feet apart from centre to centre, and are nine in number, in addition to one across each hipped end. The hip trusses are of different design, as shown in fig. 749. The roof was in good condition at the time of the writer's visit in 1901.

Fig. 4.—One of the principals of the roof of the East parish church, Aberdeen. There are five principal trusses, placed 14 feet apart. The following are the dimensions of the timbers:—

A, Tie-beam, in two thicknesses, 14 × 10 inches.
C, Principal rafters, 13 inches deep at bottom, $11\frac{1}{2}$ inches at top, and $10\frac{1}{2}$ inches thick. The rafters bear on oak abutment pieces 11 × $7\frac{1}{2}$ inches, bolted between the ties and to each other.
D, Collar-beam, in two thicknesses, one on each side of the rafter, and notched and bolted, 12 × $5\frac{1}{2}$ inches each.
E, Purlins. The two lower, 13 × $6\frac{1}{2}$ inches; the upper, $11\frac{1}{2}$ × $8\frac{1}{2}$ inches; notched on the rafters and bolted.
F, Common rafters, $5\frac{1}{2}$ × $2\frac{1}{2}$ inches, and 13 inches apart.

The discharging posts between the bracket pieces and the stone corbel are of oak, 6 inches square. Binding pieces, $9\frac{1}{2}$ × $3\frac{1}{2}$ inches, extend between the tie-beams, and are mortised into them; and into these binding pieces the ceiling joists, which are 13 inches apart, and 6 × $1\frac{3}{4}$ inches, are mortised.

The dimensions of the ironwork are as follows:—

King-rod, $1\frac{3}{4}$ inch square, with a cast-iron key-piece at top.
Queen-rods, $1\frac{1}{2}$ inch square, having solid heads at rafters, and secured at foot by being passed through solid oak pieces *k* (placed between the flitches of the tie-beams and securely bolted), and there fastened with cast-iron washers and nuts.
Four bolts at abutment end of ties ... ... $\frac{7}{8}$ inch square.
Two bolts at each oak piece, for suspending rods ... ... $\frac{7}{8}$ " "
Two bolts at each end of collar-beam ... ... $\frac{7}{8}$ " "
Purlin bolts ... ... $\frac{3}{4}$ " "

The abutments of the rafters at both ends, and the bearings of the bolts, have pieces of milled lead interposed; and all the joints of the framing were coated with white-lead and oil before being put together.

Fig. 750 shows a compound truss adapted to a hall or church with nave and aisles. The framing is simple and good: A, principal tie; B, tie of aisle roof; C, girder supported by the iron column D; and E, story-post.

Fig. 751 is a queen-post roof, adapted to the same use as the last.

The truss illustrated in fig. 752 contains a curved rib C C, formed of two thicknesses

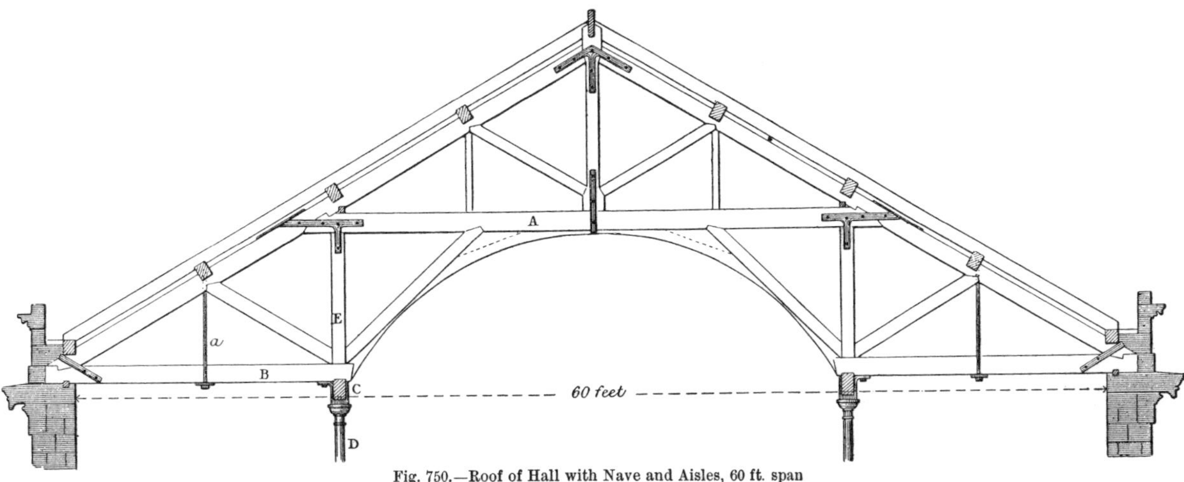

Fig. 750.—Roof of Hall with Nave and Aisles, 60 ft. span

of 2-inch plank bolted together. Its ends are let into the tie-beam; and it is also firmly braced to the tie-beam by the king-post and suspending pieces B B, which are in two

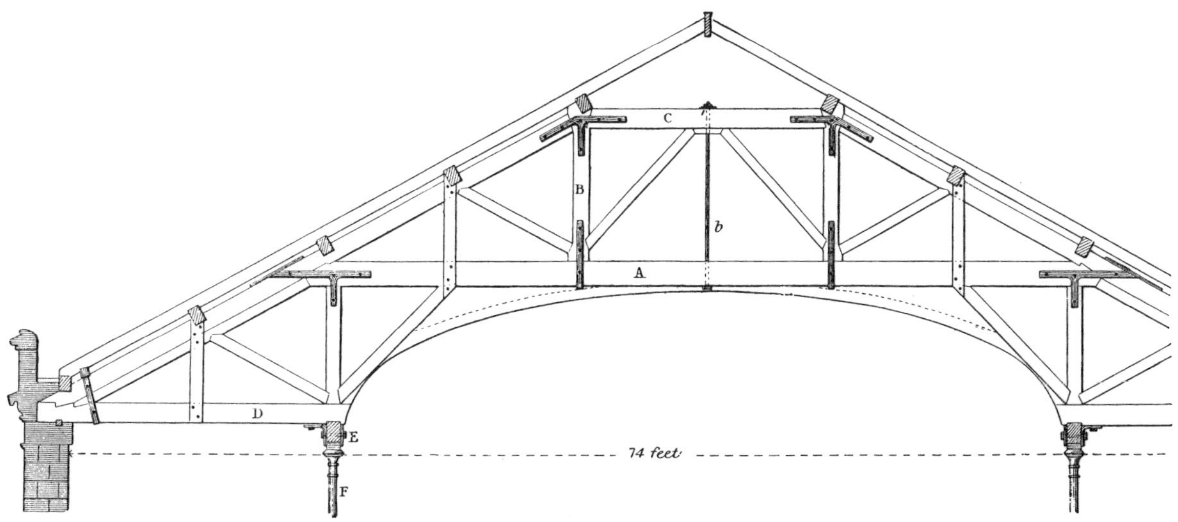

Fig. 751.—Roof of Hall with Nave and Aisles, 74 ft. span

thicknesses, one on each side of the rib and tie-beam, and by the straps *a a*. A is the rafter; *d* the gutter bearer; *c* and *b* the straps of the king-post. The second purlins are carried by the upper ends of the suspending pieces B B.

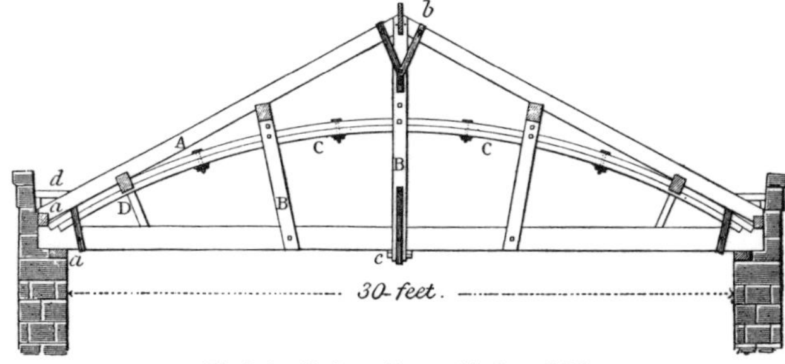

Fig. 752.—Tie-beam Truss with Curved Rib

The system of trussing designed by Mr. H. P. Holt of Manchester is well worth mention. The details are patented, and a large number of roofs on this system have been erected by Messrs. Handyside & Co. of Derby. Different designs are used for spans ranging from 20 to 50 feet, two examples being given in Plate XXX. In the first (fig. 1) the span is 30 feet, and two struts are used to support each principal rafter. In the second (fig. 2) the span is 50 feet, and each rafter is supported by three struts. The trusses are calculated for 10-feet bays. The scantlings are figured on the drawings, and are smaller than those usually adopted. The ironwork is cleverly designed, and is fully shown in the details given in the plate.

The truss illustrated in fig. 753 represents one of two pitch-pine trusses erected from the writer's designs over a board school. The span is 30 feet, and the slope 45°. The

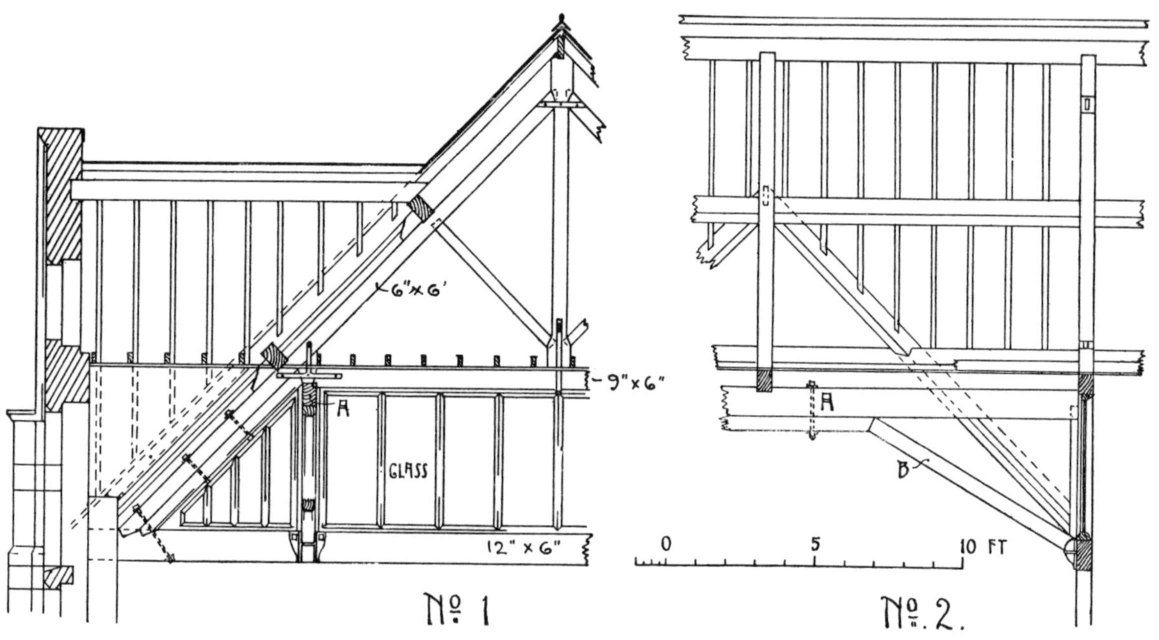

Fig. 753.—Compound Truss with Glazed Panels

queen-posts are 3 inches thicker than the tie-beams and straining-beams, and serve to support two longitudinal timbers A, on which a king-post truss is carried; these bearers are stiffened by the struts B. To the tie-beam of each of the main trusses the top-rail of a folding partition is bolted, and the spaces between the tie-beams and collars are divided into panels by rails and bars, the panels being glazed. The lower parts of the principal rafters consist of two timbers bolted together.

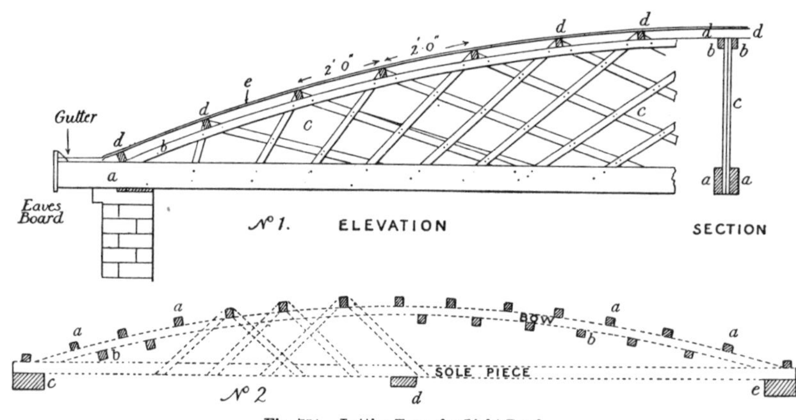

Fig. 754.—Lattice Truss for Light Roofs

For the roofs of temporary sheds or workshops light lattice trusses are now often used, as shown in fig. 754, No. 1. These consist of tie-beams or sole-pieces *a a*, bow-pieces *b b*, and lattices *c c*, and support purlins *d d*, to which the sheeting *e* is nailed. Messrs. D. Anderson & Son of Belfast, the well-known manufacturers of roofing felts, give the following scantlings for roofs of this kind:—

	Span 20 feet.	Span 30 feet.	Span 40 feet.	Span 50 feet.	Span 60 feet.	Span 70 feet.	Span 80 feet.	
	inches.	inches.	inches.	inches.	inches.	inches.	inches.	
Sole-pieces (double) ...	$4\frac{1}{2} \times \frac{3}{4}$	7×1	$8 \times 1\frac{1}{8}$	$9 \times 1\frac{1}{4}$	$10 \times 1\frac{3}{8}$	$11 \times 1\frac{5}{8}$	$12 \times 1\frac{5}{8}$	Pitch-pine.
Bows (double)	$1\frac{1}{2} \times \frac{3}{4}$	2×1	$2\frac{1}{4} \times 1\frac{1}{8}$	$2\frac{1}{2} \times 1\frac{1}{4}$	$2\frac{3}{4} \times 1\frac{3}{8}$	$3 \times 1\frac{1}{2}$	$3\frac{1}{4} \times 1\frac{5}{8}$	Pitch-pine.
Lattices	$3 \times \frac{3}{4}$	$3 \times \frac{3}{4}$	$3 \times \frac{7}{8}$	$3\frac{1}{2} \times \frac{7}{8}$	$3\frac{1}{2} \times \frac{7}{8}$	$3\frac{1}{2} \times \frac{7}{8}$	4×1	Spruce.

A tongue-piece about 5 feet long, 1 inch deeper than the sole-pieces, and equal in thickness to two lattice-boards, should be fixed between the ends of the sole-pieces. The trusses should be firmly fixed to the wall-plates, and not more than 8 or 9 feet apart. The purlins are $3\frac{1}{2}$ inches by 2 inches and 2 feet apart, headed on alternate trusses and firmly nailed, and the sheeting to receive the felt is $\frac{1}{2}$ inch thick and nailed to the purlins. The rise of the truss in the centre should be $1\frac{3}{8}$ inch for every foot of the span.

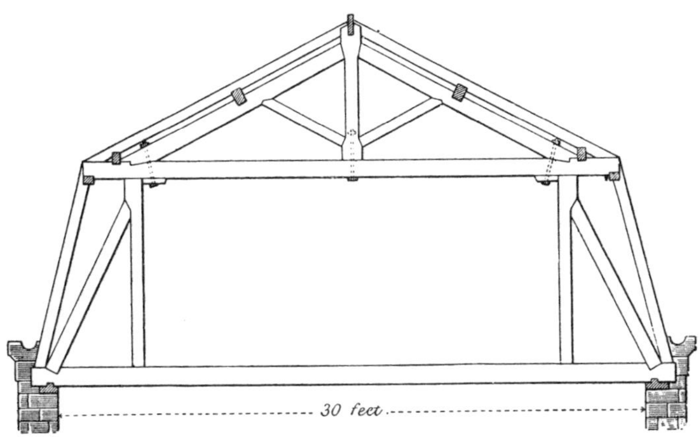

Fig. 755.—King-post Truss for Mansard Roof

To construct the truss, mark out the size on a wooden bench or floor, and nail blocks thereto as shown in No. 2, the upper ones *a a* being of the same size and in the same positions as the purlins; the lower blocks *b b* serve merely to hold the timber of the bow in position until nailed. The blocks *c* and *e* represent the two wall-plates, and the block *d* is fixed from 1 to $1\frac{1}{2}$ inch above the line joining *c* and *e* in order to give a slight camber to the sole-pieces. One sole-piece must now be laid down. Then cut the end of the bow-piece to the required angle, and after spiking it firmly to the sole bend the bow into an arc, pressing it between the blocks as shown. The lattice-boards can now be laid on and nailed to the sole-pieces and bows, and also at all intersections. Fix the tongue-pieces at the ends of the sole, and finally lay on and firmly nail the second sole- and bow-pieces.

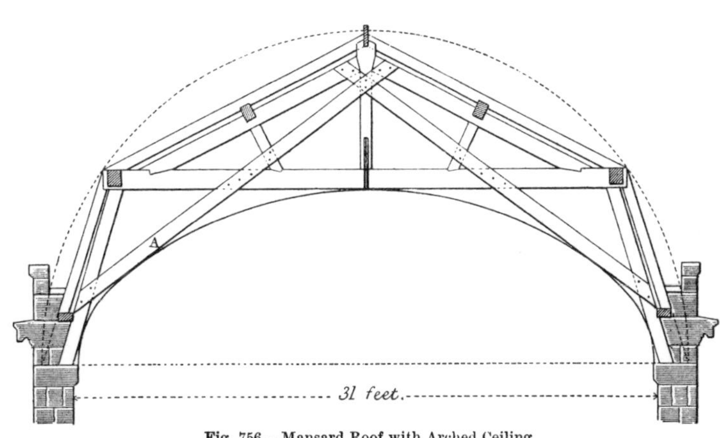

Fig. 756.—Mansard Roof with Arched Ceiling

The felt may be laid from gable to gable or from eaves to eaves, the edges being overlapped about $1\frac{1}{2}$ inch and nailed at intervals of about 2 inches with $\frac{7}{8}$ or 1-inch flat-headed tinned or galvanized tacks.

Curb or Mansard Roofs.—Fig. 755 shows a king-post Mansard roof, in which there is a wide space available as an apartment. The construction would be improved by adding straps to the feet of the story-posts. In fig. 756 two raking ties A are introduced instead of the lower tie-beam, so that space is obtained for an arched ceiling.

A queen-post Mansard roof erected over the riding-house at Copenhagen is given in fig. 757. This is regularly trussed. Colonel Emy remarks that there should be struts *a a a* under the purlins, and a small collar-piece *b* added; that the tie B is too heavy, and that the cross-piece above D serves no useful purpose, and might be dispensed with.

Mansard roofs form an economical and convenient method of construction for factories in which ample side-light is desired in the uppermost stories. The writer has designed a considerable number for clothing factories on the lines of that shown in fig. 758. The tie-beam serves to carry the floor-joists, and is supported in the middle by a cast-iron pillar. The story-posts are inclined and prevented from being thrust outwards by the cast-iron angle-pieces which are bolted to them and to the upper tie-beam. Lateral movement is prevented by the floor-joists, purlins, and upper and lower pole-plates. The steep sides of the roof between the pole-plates are entirely filled with glazed "top-lights" bolted to the plates and draining into cast-iron gutters. The roof may be ceiled at the level of the upper

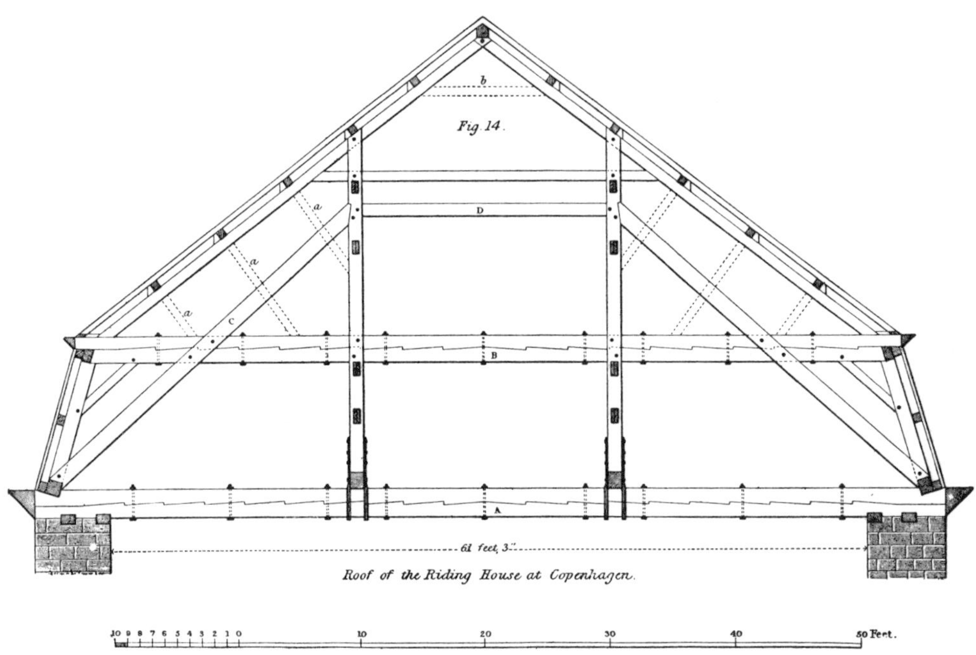

Fig. 757.—Mansard Roof of Riding-house, Copenhagen

tie-beam, or left open to the ridge. This method of construction leaves the whole of the floor-space free from pillars or other supports, and is therefore in many cases preferable to that shown in fig. 755.

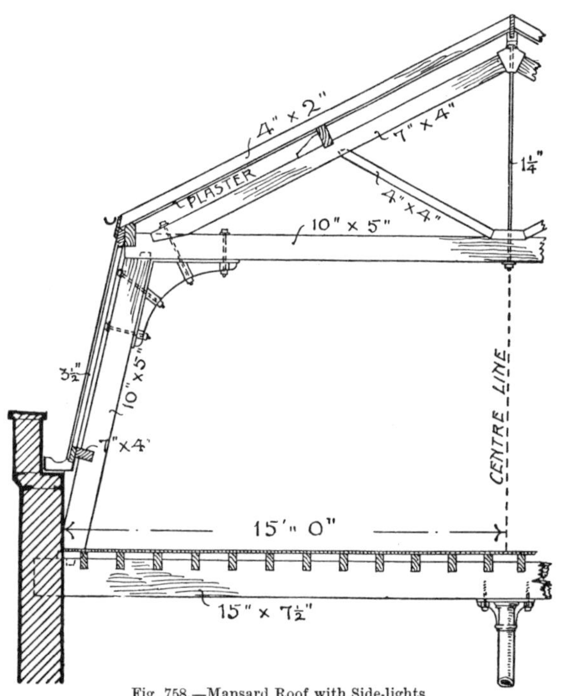

Fig. 758.—Mansard Roof with Side-lights

6. ROOF-PRINCIPALS WITHOUT TIE-BEAMS

1. *The Collar-beam Truss.*—Probably no kind of truss has been more frequently adopted for the roofs of board schools than this. It is most suitable for roofs of moderate width, of which the rise exceeds one-third of the span. The principal advantage is that additional height and air-space can be given to the rooms by forming the ceilings at the level of the collars. The principal objection is that the truss tends to spread and thrust out the walls. Buttresses are often built to counteract this thrust, and curved braces are usually introduced for the same purpose. Sometimes cast-iron segments are bolted to the collars and rafters in the manner shown in fig. 758.

In its simplest form the truss consists of two principal rafters and a collar-beam, arranged as in fig. 724, page 19. Straps should be provided to secure the ends of the collar to the principal rafters, or the collar may be in two flitches, one on each side of the rafters and bolted to them.

A typical mediæval example of a simple collar-beam roof is given in fig. 759. Arched braces are let into the rafters and collar by tongue-and-groove joints secured with trenails.

The ends of the collars are also fixed with trenails to the rafters. Arched wind-braces are

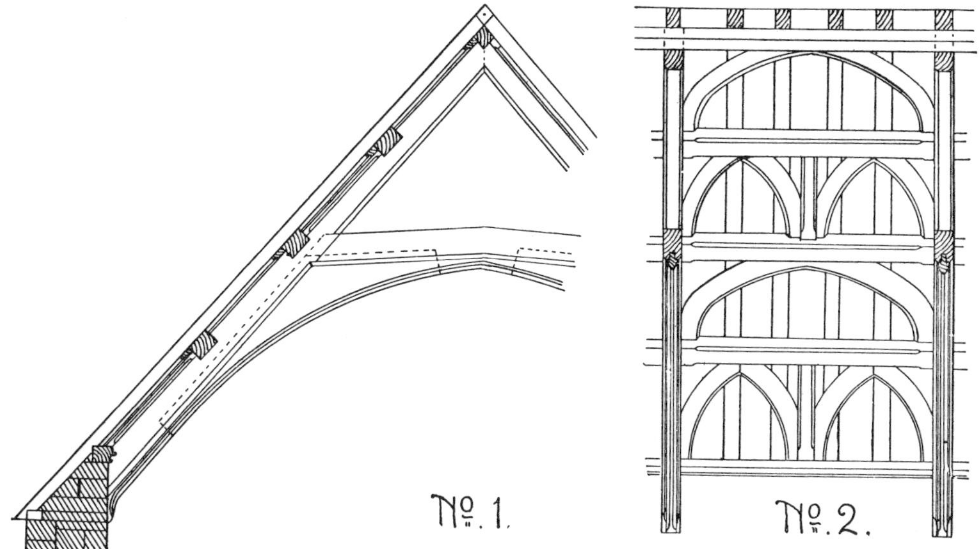

Fig. 759.—Roof of Hall at Martock, Somersetshire

introduced to support the purlins, as shown in the longitudinal section of the roof. The span of the roof is 22 feet 6 inches, the collars 12 × 7 inches, and the curved ribs $3\frac{1}{4}$ inches thick.

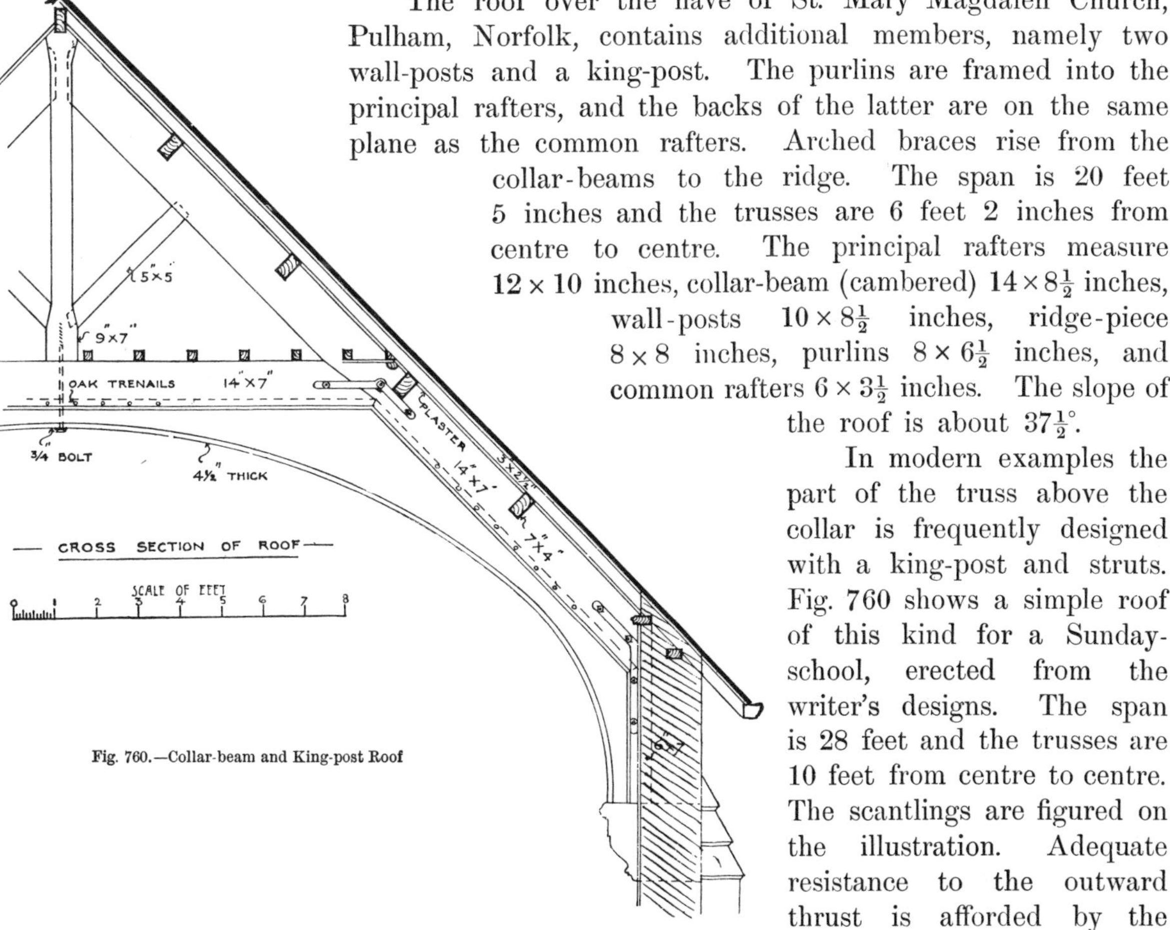

Fig. 760.—Collar-beam and King-post Roof

The roof over the nave of St. Mary Magdalen Church, Pulham, Norfolk, contains additional members, namely two wall-posts and a king-post. The purlins are framed into the principal rafters, and the backs of the latter are on the same plane as the common rafters. Arched braces rise from the collar-beams to the ridge. The span is 20 feet 5 inches and the trusses are 6 feet 2 inches from centre to centre. The principal rafters measure 12 × 10 inches, collar-beam (cambered) $14 \times 8\frac{1}{2}$ inches, wall-posts $10 \times 8\frac{1}{2}$ inches, ridge-piece 8 × 8 inches, purlins $8 \times 6\frac{1}{2}$ inches, and common rafters $6 \times 3\frac{1}{2}$ inches. The slope of the roof is about $37\frac{1}{2}°$.

In modern examples the part of the truss above the collar is frequently designed with a king-post and struts. Fig. 760 shows a simple roof of this kind for a Sunday-school, erected from the writer's designs. The span is 28 feet and the trusses are 10 feet from centre to centre. The scantlings are figured on the illustration. Adequate resistance to the outward thrust is afforded by the masonry buttresses.

A collar-beam roof with tie-rods and king-rod, erected over one of the sheds at Liverpool Docks, is shown in fig. 761. The timbers are very heavy, but in the absence of information as to the width of the bays it is impossible to say whether they are unduly heavy or not. The principal rafters are 14 × 8 inches, collar-beams in two 11 × 3-inch flitches (one on each side of the rafters), purlins 16 × 4 inches, and tie-rods and king-rod $1\frac{3}{4}$ inch in diameter. The purlins are framed into the rafters.

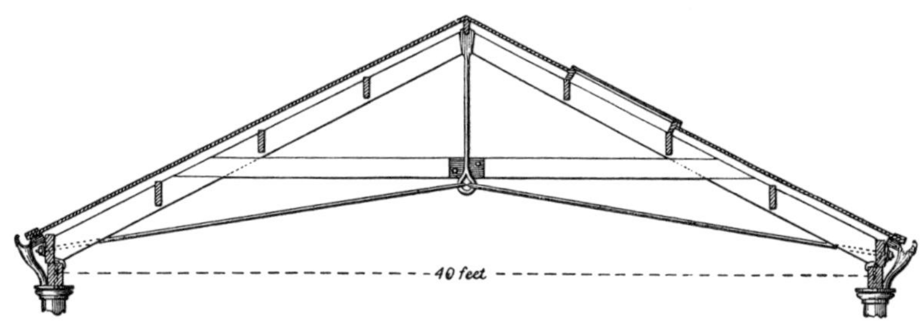

Fig. 761.—Collar-beam Truss with Tie-rods, over Shed at Liverpool Docks

The tie-rods are sometimes arranged like those in a scissor-beam truss (fig. 553, page 325, Vol. I), each rod extending from the foot of one rafter to the back of the other rafter at its junction with the collar. An example on these lines with wood ties will be found in fig. 766. Indeed nearly every form of trussed-rafter roof is adapted for principals, if the scantlings are increased.

Still another form of collar-beam truss, adapted for larger spans, is shown in fig. 762. The collar-beam is composed of two flitches, one on each side of the rafters and suspending posts. Two flat wrought-iron ties extend from the foot of each queen-post to the foot of the nearer rafter, and are secured to these members by bolts. The part of the truss above the collars is framed as a compound king-post truss.

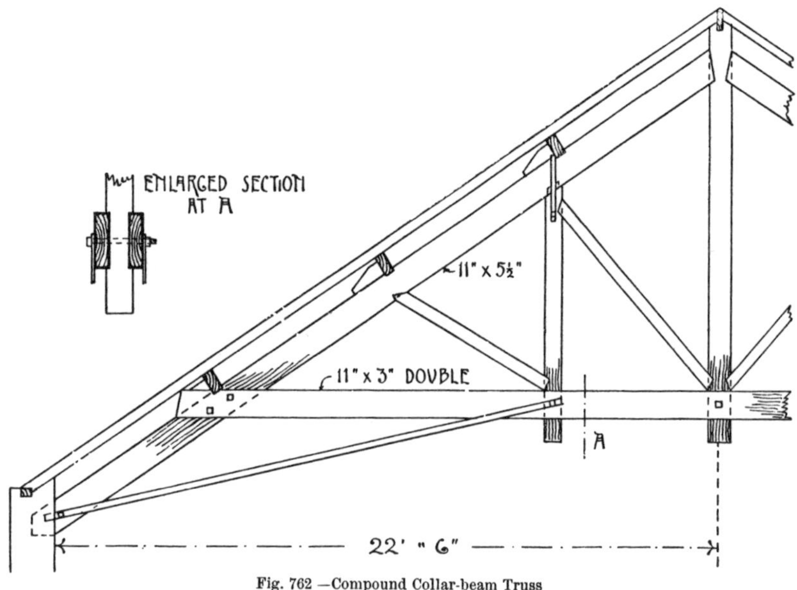

Fig. 762.—Compound Collar-beam Truss

2. *The Hammer-beam Truss.*—The hammer-beam truss is in many cases merely a variety of the collar-beam truss. The roof of the Great Hall, Penshurst Place, Kent (fig. 763), shows a rudimentary hammer-beam at A, supported by the carved oak figure below. This roof is an interesting example of fourteenth-century design. The span is 38 feet 9 inches, and the trusses are about 16 feet from centre to centre. The upper parts of all the common rafters, as well as of the principal rafters, are trussed. The main trusses serve to support three longitudinal timbers, two of which are shown in the illustrations, and these in turn support the trussed rafters above, short "puncheons" or ashlaring being introduced above the lower timbers for this purpose. The arched braces are 12 inches thick, collar-beam 16 inches thick and about 18 inches deep at the centre of the span, moulded purlins 14 × $14\frac{1}{2}$ inches, upper longitudinal timber 9 × 9 inches, and common rafters about 9 × $7\frac{1}{2}$ inches. This illustration is taken from Dollman and Jobbings' *Ancient Domestic Architecture in Great Britain.*[1]

The next illustration (fig. 764) is from the same valuable work, and shows the hammer-beam fully developed. It represents the roof of the hall of Colston's House, Small Street, Bristol, dating probably from the fifteenth century. The span is only 18 feet 5 inches, and the bays about 8 feet. All the members of the truss, with the exception of the hammer-

[1] B. T. Batsford, London.

ROOF OF THE GREAT HALL, HAMPTON COURT

beams and wall-posts, are $5\frac{1}{2}$ inches thick. Arched wind-braces are introduced between the two purlins and between the lower purlin and the cornice or wall-plate.

In some cases two heights of hammer-beams are introduced, and in others the collars are omitted, the arched braces extending from the free ends of the hammer-beams to the ridge. In modern examples a tie-rod is sometimes introduced to connect the two hammer-

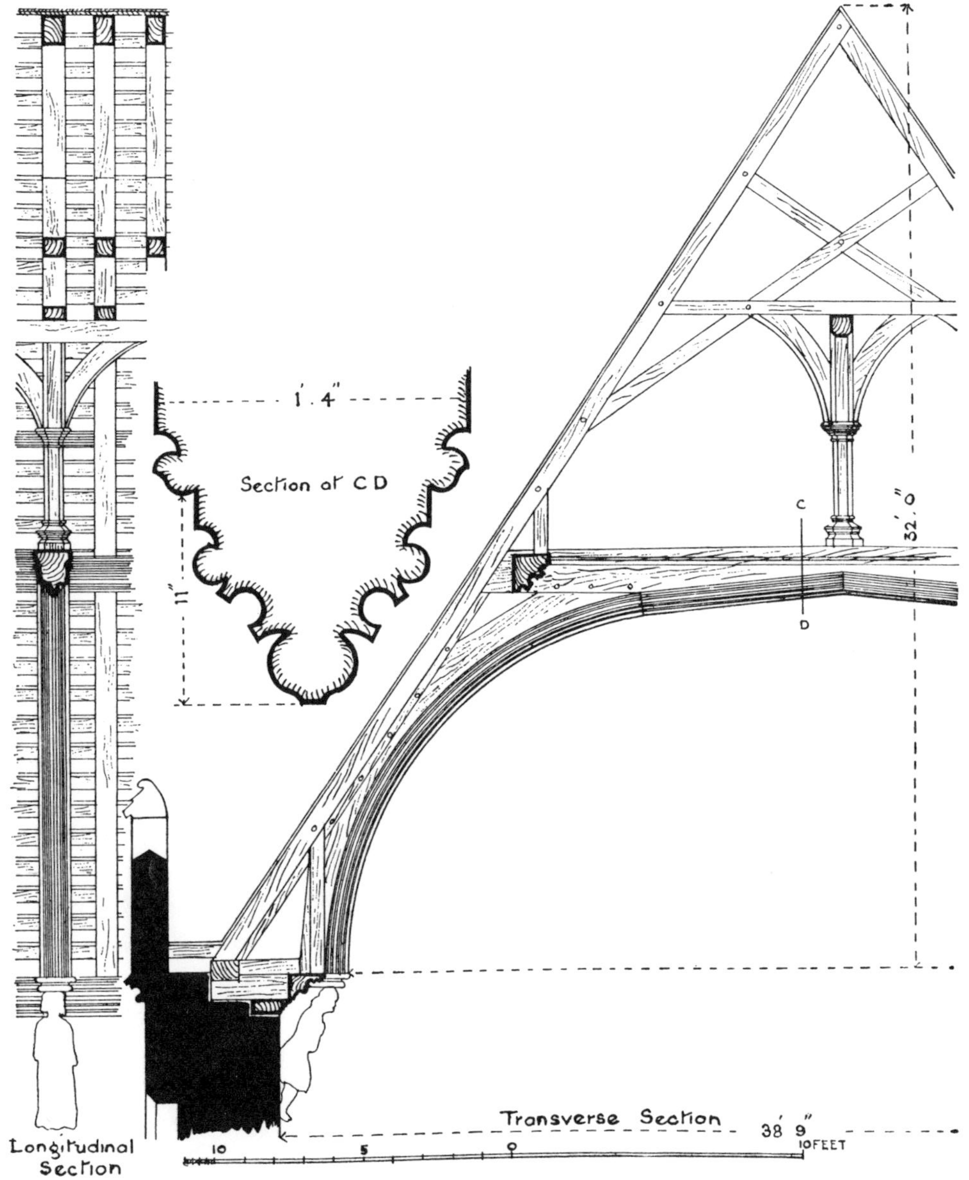

Fig 763.—Roof of the Great Hall, Penshurst Place, Kent

beams, thus converting the truss in reality into a tie-beam truss, although the hammer-beam appearance is retained. The fine roof over the Whitworth Hall, Owens College, Manchester, designed by Messrs. Alfred Waterhouse, R. A., & Sons, is an example in point. Such a tie-rod reduces the stresses in many of the timbers, as explained on page 326, Vol. I, and also prevents the outward thrust on the walls, which many hammer-beam roofs otherwise exert. Figs. 1 and 2, Plate XXXV, illustrate a truss with a timber tie-beam supported on curved brackets, which give to it the appearance of a hammer-beam truss.

Plate XXXI.—Fig. 1 is a longitudinal, and fig. 2 a transverse section of the roof of the great hall, Hampton Court. The hall is 106 feet long, 40 feet wide, and 45 feet high in the walls. It was completed in 1536 or 1537. The roof consists of seven bays in length, one of which is the subject of fig. 1, and by referring to the transverse section (fig. 2) the construction will be clearly comprehended. Each principal consists of a centre arch and two half-arches, and the principals are connected longitudinally by three tiers of arches, as seen in fig. 1. These, with their enriched panels and pendants, produce a rich effect.

Fig. 3 is one of the pendants of the second tier of arches; fig. 4 one of the large pendants; fig. 6 one of the pendants of the third tier of arches; figs. 7 and 8 are two of the wall-corbels from which the roof springs. The exterior of the roof has a double pitch like the Mansard roof.

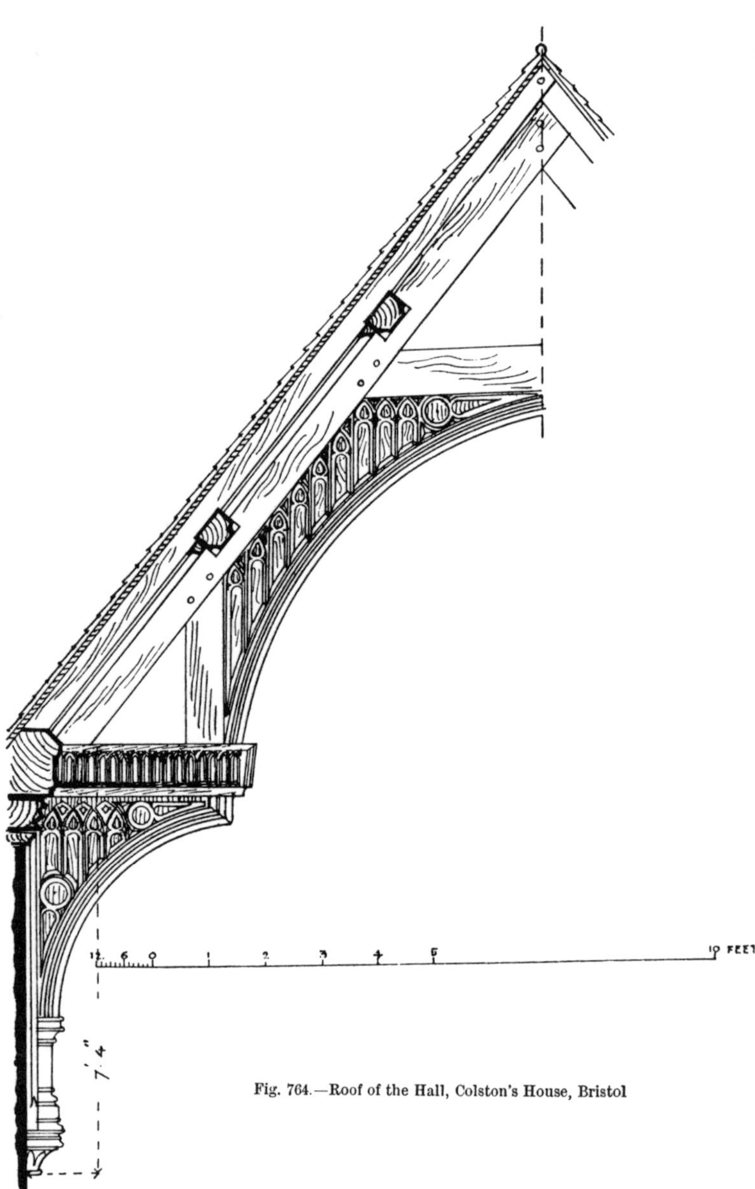

Fig. 764.—Roof of the Hall, Colston's House, Bristol

The figures in Plate XXXII represent the roof of Westminster Hall, which was rebuilt by Richard II near the close of the fourteenth century. The dormers were inserted at a later date, and have been removed since this plate was prepared. The drawings are not strictly accurate, but give a good general representation of this fine example of carpentry.

In fig. 1 is shown the half of one of the principals. It is composed of two rafters *a a*, joining at the apex on a puncheon, and united in the middle of their length by a collar-beam *d*. There is no tie-beam, but in its place two hammer-beams *e*, one at each side, which receive the ends of the rafters. These hammer-beams are horizontal, and are sustained by the arc or bracket *f*. The end of each hammer-beam sustains a post *g*, decorated with a base and capital, and into this post is framed the rafter or strut *c*, which discharges part of the weight, and is supported at the foot on a block *h*, lying on the inner end of the hammer-beam. It is prolonged below the hammer-beam at *i*, and abuts against a stile placed on the wall. Between the collar-beam and the ridge there is another beam *k*, framed into the posts *n*, which sustain the rafters at the point where the purlins *o* rest. A great arc *m* springs from the same corbel as the bracket of the hammer-beam at the middle of the height of the walls. This arc is in two pieces, and embraces in its thickness the posts *g* and the collar-beam *d*. A smaller arc *r* springs from the end of the hammer-beam and joins the great arc. This discharges part of the load to the point of the hammer-beam, whence it is carried by the lower arc to the corbel. These parts are all connected and firmly braced together, and

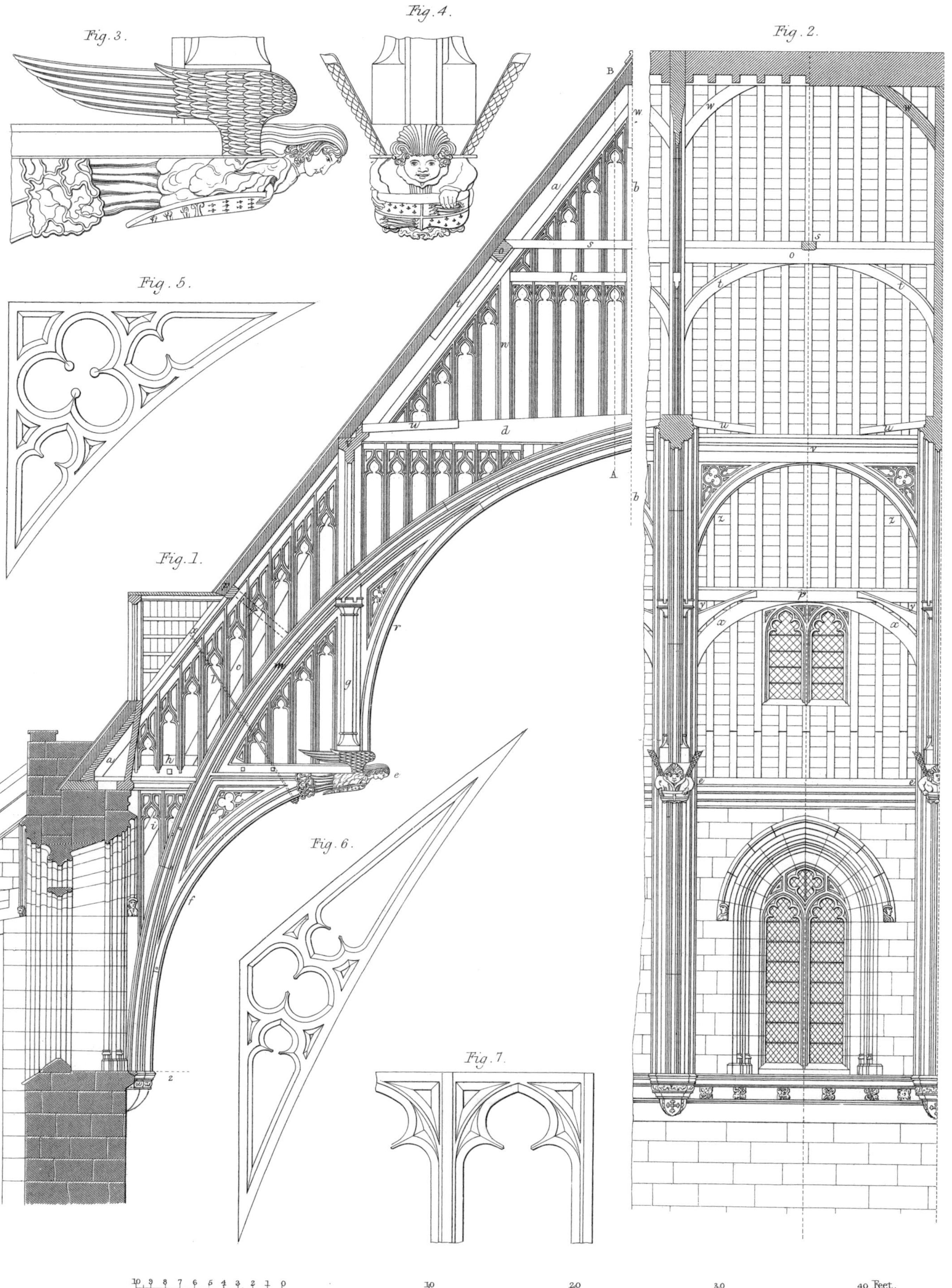

ROOF OF WESTMINSTER HALL

to the rafter, by the iron rod l, which passes through them, and is secured by a screw and nut at the under side of the last-mentioned arc under the hammer-beam.

Fig. 2 is a longitudinal section of one bay of the roof comprised between two principals. The roof, it will be seen, is in three divisions: the first, rising from the cornice, terminates at the hammer-beam; the second extends from the hammer-beam to the collar-beam; and the third from that to the ridge. Curved pieces *w t x* are introduced to support the timbers longitudinally; and a collar-beam is introduced between the common rafters to prevent them sagging under the weight of the covering, and they are further strutted by the pieces *u u*. The lower purlins *p* of the roof are also sustained by two struts *y* abutting against the great arc.

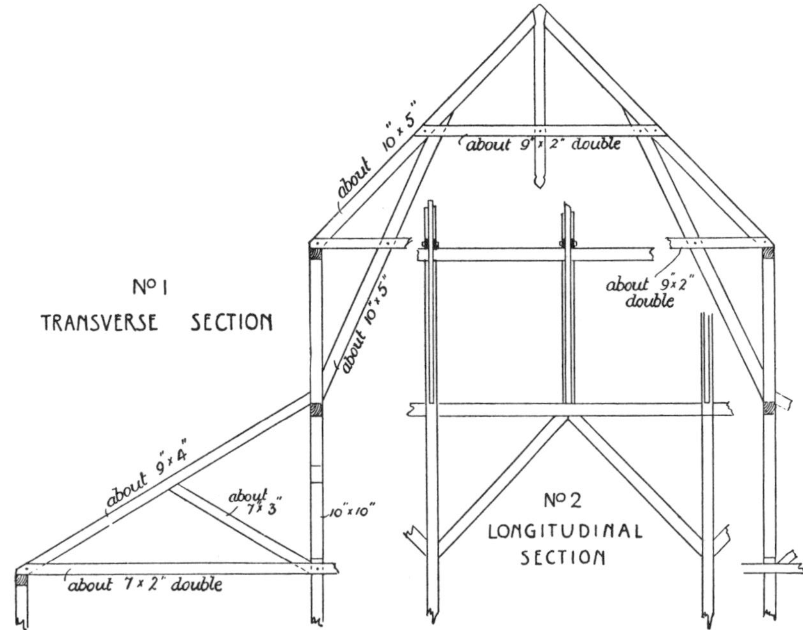

Fig. 765.[1]—Roof of Russian Agriculture Pavilion, Glasgow Exhibition

Figs. 3 and 4 are representations, to a larger scale, of the angels sculptured at the extremity of the hammer-beams. Each angel holds a shield with the arms of England and France quartered. Fig. 5 is the detail of the spandrel at *z z*, fig. 2. Fig. 6 is the spandrel at *r*, fig. 1. The tracery above the collar-beam is shown in fig. 7.

The hammer-beam truss is also used for strictly utilitarian structures, but more frequently abroad than in this country. Two examples, designed by Zelenko, an architect of Moscow, appeared in the Russian Pavilions of the Glasgow Exhibition of 1901. That shown in fig. 765 formed the roof of the pavilion devoted to agriculture. The span of the central portion or nave is about 33 feet, and that of each side portion about 21 feet. The roofs are supported on posts 10 inches square and about 20 feet apart; these posts form part of the alternate trusses, but other trusses are supported (midway between the posts longitudinally) by beams framed

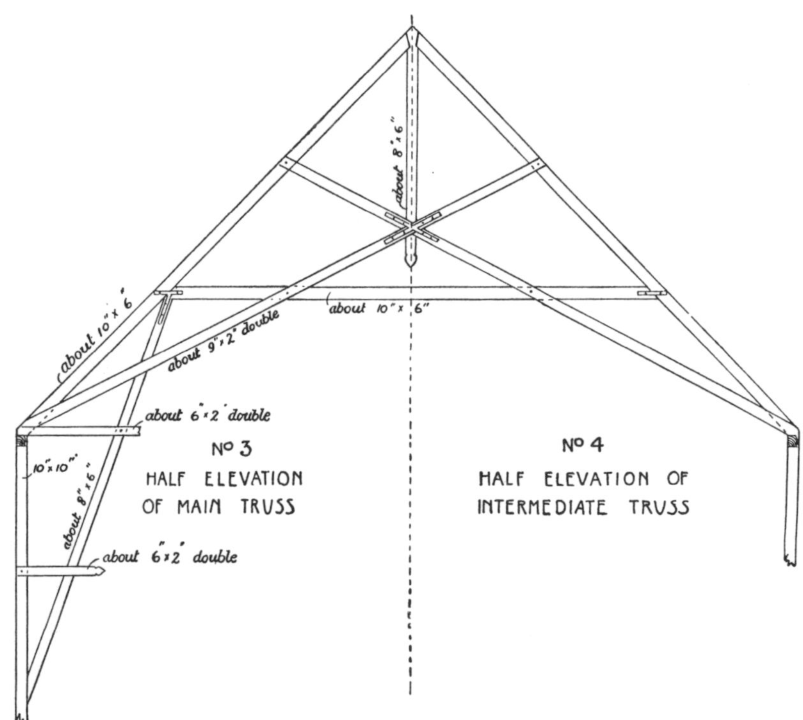

Fig. 766.[1]—Roof of Russian Forestry Building, Glasgow Exhibition

[1] These illustrations appeared in *The Builder*, Sept. 14, 1901.

into the posts and stiffened by struts. The collars and hammer-beams are composed of two flitches, bolted to the other timbers at the intersections. The drawing is not a strictly accurate representation of the roof, as the writer prepared it from his sketches and notes, and not from detailed measurements, but it serves to illustrate the general character of the framing.

The next example (fig. 766), to which the last remarks also apply, is the roof of the Forestry Building. This is of more intricate design, the span being about 60 feet. Here again the supporting posts are 10 inches square, and form part of the main trusses (No. 3). These have two heights of hammer-beams, and are about 17 feet apart. Midway between them, intermediate trusses of simpler design (No. 4) are fixed, approximating to the German truss illustrated in fig. 729. The hammer-beams and scissor-beams are composed of two flitches. T-shaped straps are provided to connect the ends of the collars to the principal rafters, and cross-straps at the intersection of the scissor-beams. Ordinary bolts are used at the other intersections.

The roof illustrated in Plate XXXIII has two kinds of trusses, one of which (fig. 1) has hammer-beams A; the other is of simpler character. It is a temporary structure erected over the nave of Drontheim (Trondjhem) Cathedral in Norway to facilitate the work of rebuilding the piers and arches, and was sketched by the writer in 1900. The main trusses are on the cantilever principle, nearly the whole of the weight being carried by the compound wood posts. These posts also serve as supports for temporary galleries (not shown in the illustration) a few feet below the springing of the struts. The triangle at the apex of the truss forms a simple couple, the principal rafters being in compression and the horizontal member in tension. The intermediate trusses are supported by the walls and by longitudinal bearers, notched into the posts and stiffened by raking struts, as shown in the longitudinal section.

The purlins are about 3 feet 3 inches from centre to centre, and the roof-boarding, which runs from eaves to ridge, is nailed to them and covered with tarred paper or felt. Rolls are nailed to the boarding about 3 feet apart to form the side joints in the covering, as in the case of lead roofs. The span of the roof, the width of the bays, and the scantlings of the principal timbers are figured on the drawings. The pitch of the roof is about 15°, or about one-eighth of the span.

3. *Curved and Polygonal Trusses.*—In 1561 Philibert de Lorme published his book, entitled *New Inventions for Building Well at Little Expense.* In his address to the reader he says that as it is difficult to find trees large enough to serve for beams and the other timbers of mansions, he has long sought for some invention which would enable him to use all kinds of wood, and even the small pieces, and so to dispense with the great trees hitherto used. The result of his researches was the system of framing to which the name of the inventor is given, and which is here illustrated and described in detail.

The system of Philibert de Lorme consists in the use of arcs or hemicycles formed of planks as substitutes for the framed principal. The advantage of the system, according to its author, is the saving of expense, because very light and short timbers are proper for the work, and the walls need not be so thick as for heavier carpentry; the wood is more easily hauled and raised; and in countries where only small scantlings of timber are obtainable, the system permits of roofs of greater span being made than would otherwise be possible.

The mode of construction was first employed by De Lorme in roofing the pavilions of the Château de la Muette, at St. Germains-en-Laye. The walls of these pavilions were in a defective state, and would not bear the weight either of stone vaulting, or of heavy carpentry, even if trees large enough to make the roof of the ordinary construction in use at that time could have been obtained, which, we learn from the work of De Lorme, was not the case.

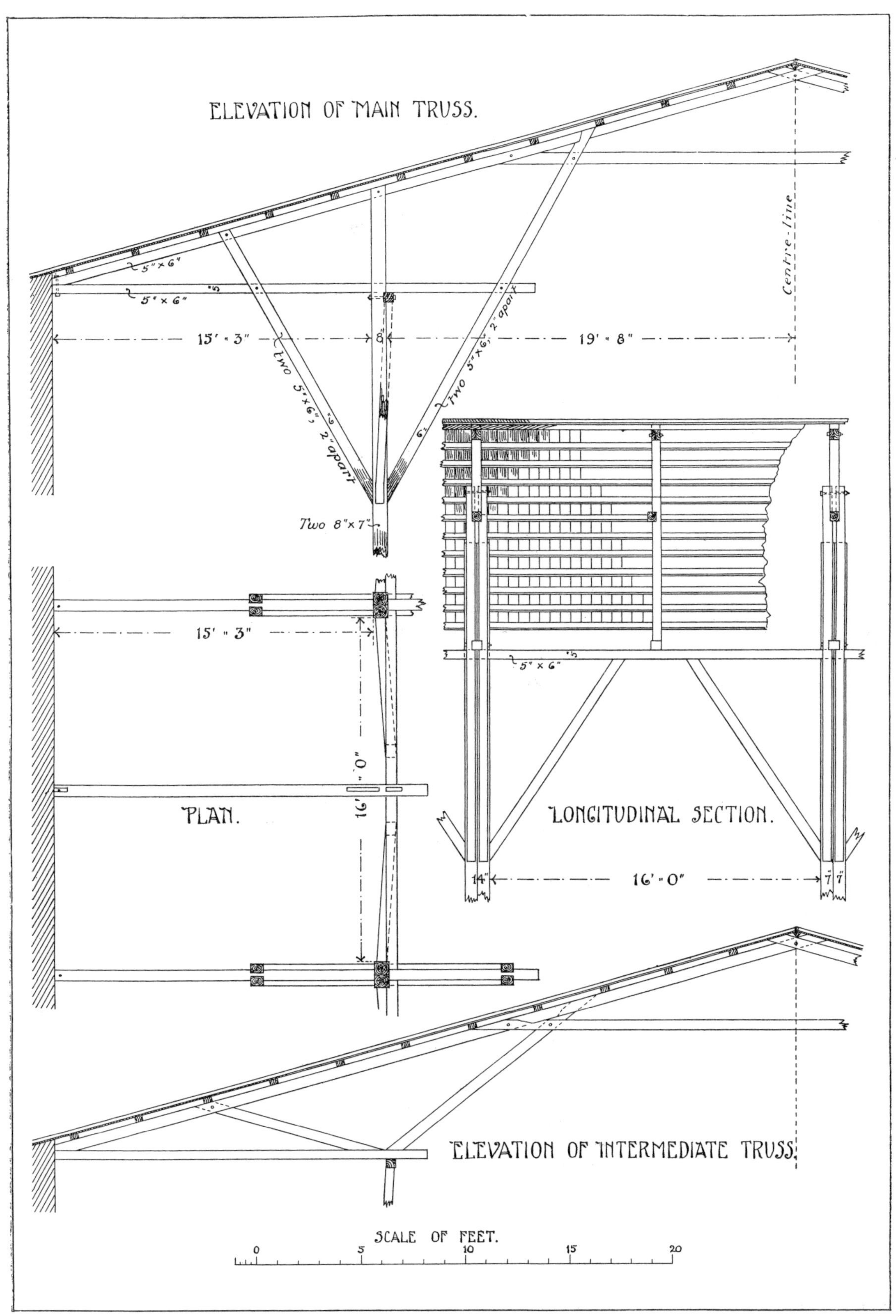

TEMPORARY ROOF OF CATHEDRAL AT DRONTHEIM (TRONDJHEM), NORWAY

Fig. 767 shows portions of two of the hemicycles, as he called his frames, for spans of from 24 to 30 feet. Each hemicycle A B is in this case built of two thicknesses of wood, each of which, *e*, *f*, is in pieces of 3 or 4 feet long, 8 inches wide, and 1 inch thick. The joints of the one series are made to fall on the middle of the length of the other; each piece has a mortise cut through it in the middle of its length, and a half-mortise or groove at each end. The mortises are 4 inches long, and a little more than 1 inch wide. They serve to receive ties *g g*, which may be of any length, and otherwise of the same dimensions as the mortise. The ties are secured in their places by keys *h h*, driven through mortises made in the ties, one on each side of every hemicycle. The mortises in the ties are made with a little *draw*. The keys are best when made of split wood. The ties and keys serve to keep the hemicycles the proper distance apart—usually about 2 feet. The two thicknesses of timber in each hemicycle are first framed together by small pins, to prevent their sliding, and then the hemicycles are united by their ties, and the two thicknesses of timber in each fastened by the keys. The hemicycles are tenoned into wall-plates, 10 or 12 inches wide and 8 or 9 inches thick, having mortises sunk 2 feet apart to receive the ends of the hemicycles. The mortises are 2 inches wide, 3 inches deep, and 6 inches long.

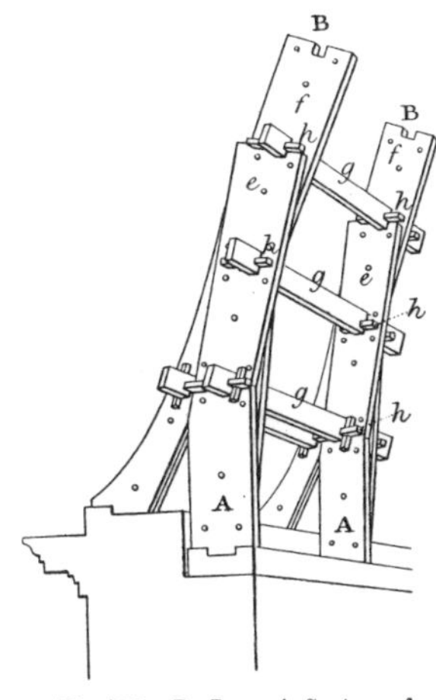

Fig. 767.—De Lorme's System of Roofing for Spans of 24 to 30 feet

In the roof of the pavilion of the Château de la Muette, where the span was 64 feet, the scantling was increased to 13 inches wide and $1\frac{1}{2}$ inch thick. The ties were alternately double and single, and were 3 inches by $1\frac{1}{2}$ inch. Each hemicycle was double-tenoned into the wall-plate. The general elevation of the roof is shown in fig. 768, together with details

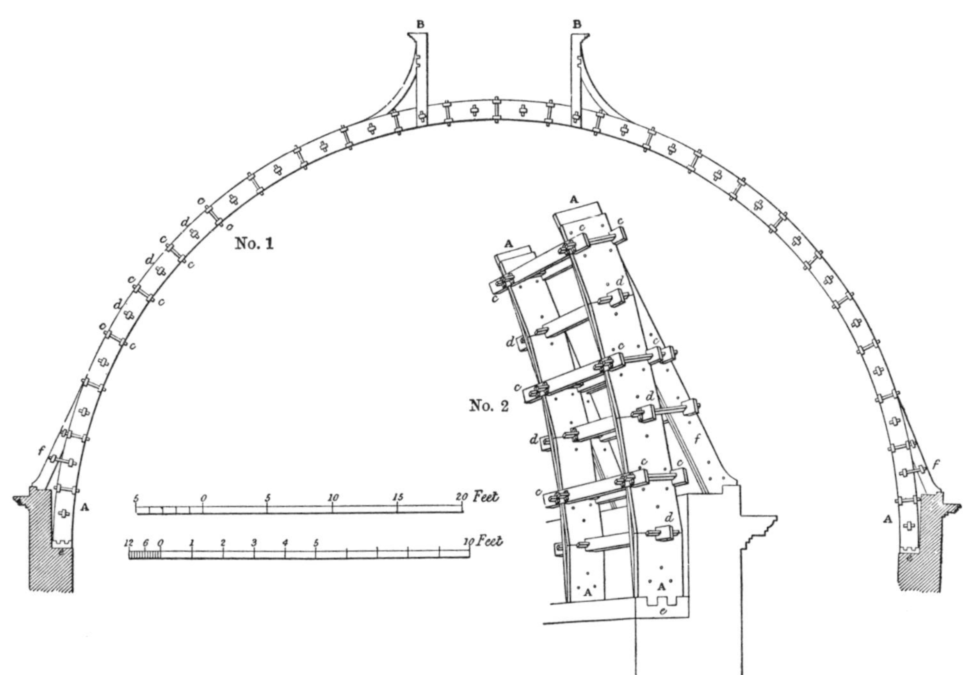

Fig. 768.—Roof of Pavilion, Château de la Muette (De Lorme's System)

of the springing to a larger scale. The same letters refer to the same parts in both drawings —A A, one of the hemicycles; B, a terrace or gallery, used as a belvidere; *c c*, double ties; *d d*, single ties; *e e*, wall-plates; *f f*, eaves-rafters. The notches for the double ties are just so deep that the outside surface of the tie is flush with the edge of the hemicycle.

When the span is small, and the curve of the roof is so quick that it becomes impossible to cover it with slates or tiles, De Lorme adopts the expedient shown in fig. 769. Fig. 770 shows the application of the principle in the construction of a groined vault, with a pendant in the centre.

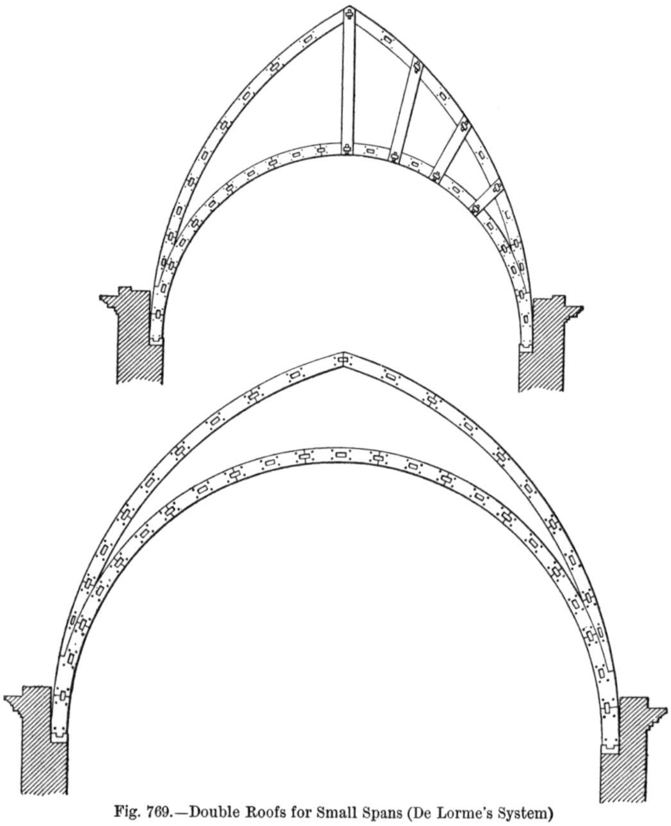

Fig. 769.—Double Roofs for Small Spans (De Lorme's System)

The dimensions of the pieces of which the arcs or hemicycles are composed, increase, of course, with the increase of the span; and, as has been mentioned above, the single ties give place to double and single ties placed alternately. In the roofs of ordinary buildings, where the span does not exceed 24 feet, the author directs the pieces which compose the hemicycles to be made 1 inch thick and 4 feet long; for roofs of 36 feet span, the thickness to be $1\frac{1}{2}$ inch; for roofs of 60 feet, the thickness to be 2 inches; for roofs of 90 feet, the thickness to be $2\frac{1}{2}$ inches; and for roofs of greater dimensions, the thickness to be 3 inches.

When Colonel Emy was called upon, in 1819, to construct the roof of a building upwards of 60 feet wide, at the barracks of Libourne, it was proposed to him to follow the method of De Lorme, but it appeared to him that where timber of tolerable length could be obtained, results equally good might be

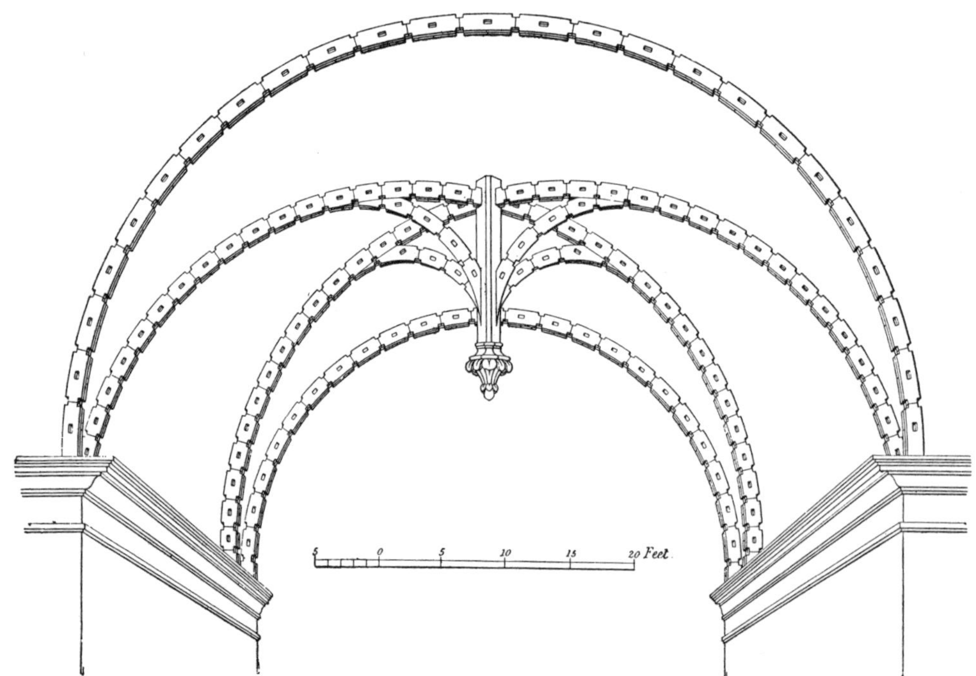

Fig. 770.—Groined Vault with Pendant (De Lorme's System)

produced without cutting it into the short scantlings required by De Lorme's system. Accordingly, as the country afforded pines of from 36 to 40 feet long, and fir still longer

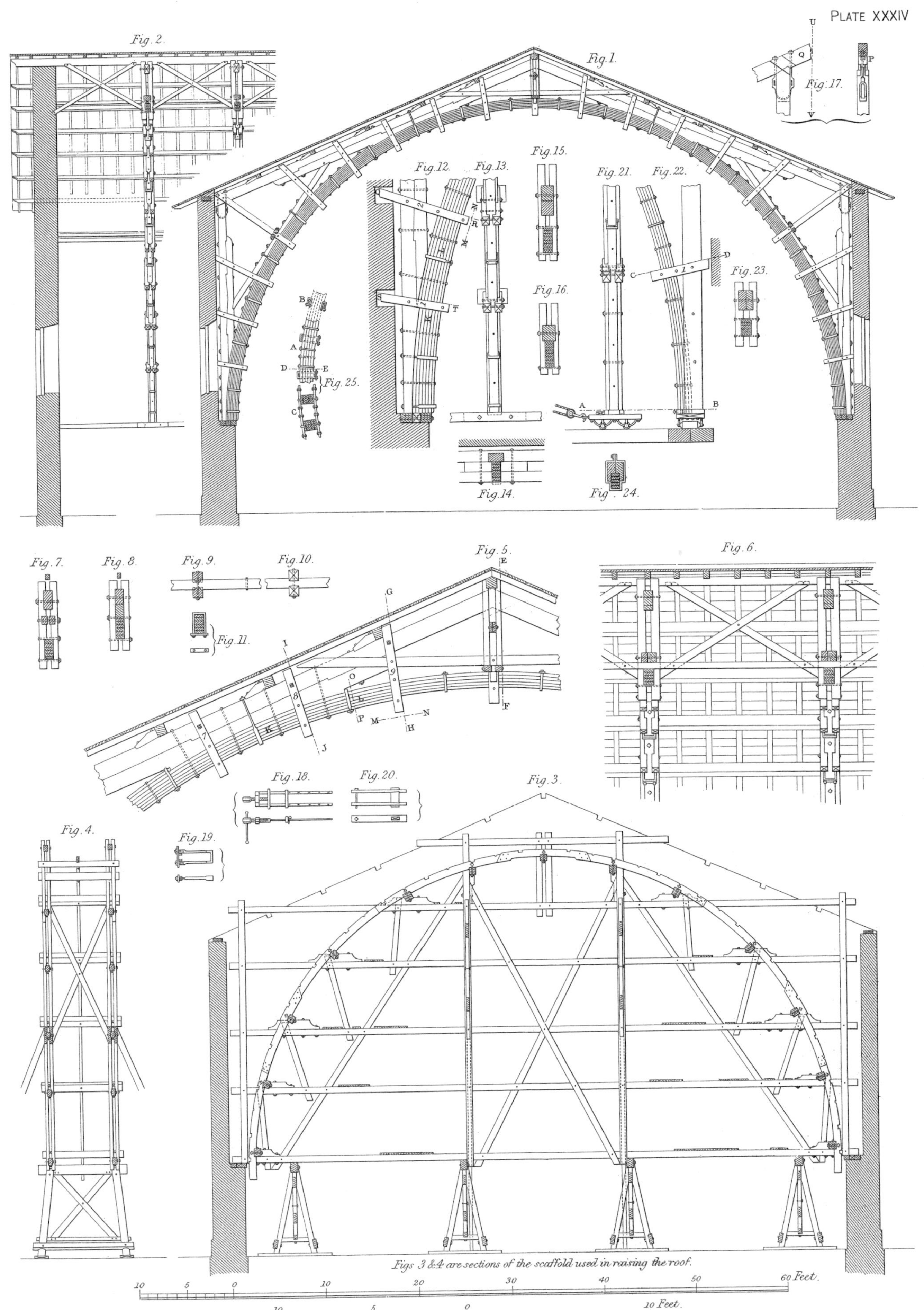

ROOF OF A SHED AT MARAC, NEAR BAYONNE, FRANCE

could be easily obtained from the Pyrenees, he sought to compose a roof in which the timber might be used in its whole length, and which should combine the necessary solidity with the lightness and economy of the system of De Lorme. He succeeded in designing a roof which, in his judgment, satisfied all those conditions; but did not obtain authority to carry out his designs. He was authorized at length to make a trial at Marac, near Bayonne, in 1825, in roofing a building of nearly the same dimensions. The success of this trial was such as, in 1826, determined the authorities to roof, in the same manner, the riding-school at Libourne, for which the system was originally designed.

The execution of M. Emy's system is within the power of the ordinary carpenter; and the workmanship is less than in the roofs of De Lorme, as the wood is all in straight pieces. There are neither mortises nor tenons, except at the ridge; and the process of construction and setting the principals in their places is so simple that at Marac, as M. Emy says, twelve workmen, two-thirds of whom were common labourers, were able to put together, and raise and set in their places, two principals each week.

Principals have often been constructed of great arcs, or centres, formed of several pieces of timber superimposed on each other. But such pieces have been of considerable scantling. They have been very short; their connection has been by iron; and their curvature produced by the adze or by heat. The construction invented by M. Emy is a timber arch composed of a series of long and thin planks applied on the flat, the flexibility of which permits them to be easily and quickly bent without the aid of heat. It would be impossible to bend, even with the aid of fire, timbers of the same scantling as these composite arcs. Even supposing that pieces of half the length could be so bent, then, to form the whole arch, butt-joints, occupying the whole section of the timber, would have to be introduced; while, by building these arch-beams of thin planks, the joints can be properly broken without seriously weakening the structure.

The combination of this system may be varied infinitely by the number, the span, and form of the arcs; and the strength of the arcs may be increased, according to the necessity of the case, without changing the system, or injuring the elegance of its appearance, by simply adding more planks either to the whole length of the arc, or to such part as trial, always indispensable in large constructions, shows to be necessary.

Since the invention of M. Emy has been made public by his own publications, and by the report of the Society for the Encouragement of National Industry, in March, 1831, roofs have been constructed on his principle with great success both for large and small spans. The examples we have engraved embrace the roof of a shed at Marac, and the roof of the riding-school at Libourne, constructed by himself (Plate XXXIV); and the application of his system to the roof of a Gothic church erected at Grassendale, near Liverpool, by Mr. Arthur Hill Holme, architect (Plate XXXV).

Plate XXXIV.—Roof of a shed at Marac, near Bayonne, France.

Fig. 1.—Transverse section of the building, and elevation of one of the principals.

Fig. 2.—Longitudinal section.

Fig. 3.—Transverse section of the building before the placing of the principals, and section of the scaffold.

Fig. 4.—Section of the scaffold, at right angles to the preceding.

Fig. 5.—Elevation of the summit of the principal in fig. 1 drawn to a larger scale.

Fig. 6.—Section through E F of fig. 5, showing the counter-bracing and ties between the principals.

Fig. 7.—Section on the line G H of fig. 5.

Fig. 8.—Section on the line I J of fig. 5.

Fig. 9.—Section of one of the radials on the line K L, fig. 5, corresponding to the longitudinal point of the arc.

Fig. 10.—End elevation on the plane M N of the radial No. 9.

Fig. 11.—Section through the arc on the line O P, showing one of the iron straps.

Fig. 12.—Side elevation of the springing of one of the arcs, showing two of the radials, Q R, S T, and the recesses, Q and S, in the wall, to receive their prolongations.

Fig. 13.—Front elevation of the same.

Fig. 14.—Portion of the plan of the wall-plates, showing the manner in which the parts of the principal are framed into them.

Fig. 15.—Section of the principal on the line of the upper face of the radial No. 2, Q R, fig. 12.

Fig. 16.—Section of the principal on the line of the upper face of the radial No. 1, S T, fig. 12.

Fig. 17.—Junction of the upright and principal rafter: Q is the elevation; P is the profile on the line U V. For the sake of distinctness the radial is not shown.

Fig. 18.—Iron strap and screw used in bending the arc plates.

Fig. 19.—One of the straps used in securing the plates to the template.

Fig. 20.—One of the ties of timber serving the same purpose.

Each principal of this roof (fig. 1) is composed of a semicircular arch, two principal vertical pieces or wall-posts, two principal rafters, two main struts from the wall-posts to the principal rafters, a king-post, and a collar-beam, the whole tied together by pieces at right angles to the curve. These radial pieces, as well as the sides of the arch, are notched upon each other. The vertical pieces are $7\frac{3}{4}$ inches thick and about 4 inches from the face of the wall. The three first radial pieces on each side are prolonged beyond the uprights, and enter recesses made in the wall to receive them, as seen in fig. 12. The object of this is merely to steady the frames, and keep them vertical.

Between the radial pieces the plates composing the arc are bolted together with cylindrical bolts, which are driven tightly into accurately-made holes by a heavy mallet. These keep the plates from sliding on each other. The plates are further firmly tied together by iron straps. The bolts are $\frac{7}{10}$ inch diameter, and about 2 feet 6 inches apart.

The plates of wood forming the arc are $1\frac{3}{8}$ inch thick, $5\frac{1}{10}$ inches broad, and about 40 feet long. Two and a half plates of this length, joined end to end, make up the whole length of the curve. The joints are so distributed that those of one row do not correspond to those of another row, and that each joint is carried by one of the radial pieces. All the plates cannot, of course, have the same number of joints. The principal rafters are $5\frac{1}{4}$ inches thick.

The principals are placed 9 feet 10 inches apart, and maintained in this position by the braces (seen in fig. 2, and on a larger scale in fig. 6), by the purlins, and by a line of double ties stretching between the fourth radial pieces.

When this roof was proposed, it was alleged that it would alter its form, and exercise a thrust upon the walls, especially when loaded with its covering. Colonel Emy, therefore, judged it proper to make several experimental principals on this construction, which could be submitted to proof, and enable him to determine what weight they could support without alteration of form, and also the number of plates of which the arcs should be composed. The experimental arcs were composed of five plates. They were sustained by thick plates of oak laid on the ground, which had first been truly levelled and beaten solid. When the arc was raised up and left to itself it drooped a little.

Then, by long cords, there were suspended to the points of the arc, which represented fairly the points of pressure, platforms of wood at about 20 inches above the ground. These platforms were then loaded gradually with cast-iron, until the weight on each reached 2200 lbs., making the total load 10·8 tons—a weight which was one quarter more than the principal would have to sustain. The plates in these experimental arcs were fastened together solely by the radial pieces and iron straps, as Colonel Emy wished to reserve a means of increasing the strength by inserting the bolts after the experiment, when the principals were set in the places they were finally to occupy.

As the weights were added, the arc appeared to flatten. At the end of twenty-four hours its curvature was tested by a radial rod of wood of 32·8 feet long mounted with iron at each end, and centred truly on an iron axis, and established with precision on the head of a pile driven in to the level of the springing of the arch. It was found that the king-post had fallen down, but that the curvature of the arc comprehended between the seventh radial pieces had not sensibly altered. There was an augmentation of the curvature below these points, the maximum being at radial No. 4; and the disposition of the principal rafters and uprights was, of course, also slightly affected. The diameter of the arc, however, did not vary; and therefore the plates must have slid on each other to the extent of not quite an eighth of an inch each.

The conclusion derived from the experiments was, that the stiffness of the arc should not be the same throughout, and that it was necessary to reinforce it in the places that had yielded the most, by supplementary plates. The proper result was obtained by adding, on the two sides of each arch, one supplementary plate to a part of the extrados, and two plates (placed end to end) to a part of the intrados. The supplementary plates were of oak, and of the same thickness as the others. The principals thus strengthened were again submitted to proof without change of form.

The following is the proportion of the number of plates, and their width, which Colonel Emy adopted as a rule:—

		Width Ft.	Width Ins.
From the springing to radial No. 1	7 plates ...	1	3
From radial No. 1 to the tie placed between radials Nos. 6 and 7 ...	8 „ ...	1	7
From the above tie to radial No. 9	6 „ ...	1	0
From radial No. 9 to king-post	5 „ (nearly)	0	11

The manner of constructing the principals was as follows:—The ground having been dressed and beaten to a level, a semicircle of 65 feet 7 inches diameter was described, representing the intrados of the arc of five plates; and the chief lines of the principal were then traced, and strong sleepers were laid down and fastened by pickets. The sleepers, twenty-four in number and 10 inches square in section, were all laid radially, and distributed so as to fall between the radial pieces of the arc and the iron straps; two only were on the outside of the springing. On these sleepers a floor was laid large enough to hold the draught of the principal; and the centre of the arc was formed by an iron axis fixed on the head of a pile. The floor being laid, and the draught made on it, pieces of wood 8 inches square were fastened through it by long spikes to the sleepers, and to these the template for curving the plates forming the arc was fixed.

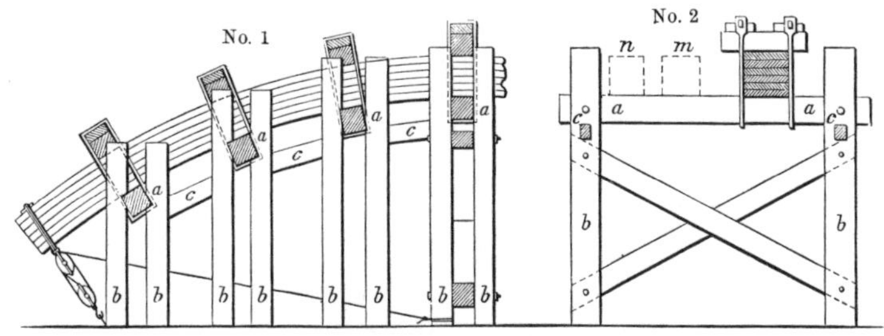

Fig. 771.—Method of Bending the Timber for the Arcs

Another method of bending the timber is also described by Colonel Emy. In fig. 771 No. 1 is a vertical projection, and No. 2 a transverse vertical section, of the apparatus. It consists of the horizontal pieces *a a*, arranged with their upper surface in the contour of the curve. They are sustained by strong framing *b b*, *c c*, *d d*. The timber is laid with its centre on the middle of the frame, and by means of purchases applied at both sides of the centre, and carried successively along to different points towards each end, it is curved, and secured by iron straps and wedges. The frame may be made wide enough to serve for the bending of other pieces, as *m*, *n*; or for a greater number, by increasing the length of the pieces *a a*, and supporting them properly.

This method is not quite perfect; for in place of the timber assuming a regular curvature, it will obviously be rather a portion of a polygonal contour. To ensure perfect regularity in the curve it is necessary to make a continuous template, in place of the several pieces *a a a a*.

It is necessary to remark that in all the cases the timber should be preserved from being injured by the iron strap by a piece of wood inserted between them, as shown in the figures. Care must be taken that the curvature given to the timber is such as will not too greatly extend, and, perhaps, rupture the fibres of the convex side, and so render it useless.

Before being bent, the timber should be softened by means of hot water or steam. A sand-bath is sometimes employed for the purpose; the apparatus is a furnace with flues, traversing the stone on which the sand is laid, in the manner of hot-house flues. There is also provided a boiler in which water is heated. On the stone a couch of sand is laid, in which the timbers are immersed, being set edgeways on a bed of sand about 6 inches thick, and having a layer of sand of the same thickness separating them, and being also covered over with sand. The fire is then lighted in the furnace, and after a time the sand is thoroughly moistened with boiling water from the boiler before mentioned. This watering is kept up all the time that the timber is in the stove. Thin deals require an hour for each inch of thickness, but for thick scantlings the time requires to be increased; for instance, a 6-inch timber should remain in the stove eight hours. Steam under pressure may also be used, but this necessitates closed chambers for the reception of the timbers.

The arc having been formed, the other parts of the principal were properly fitted, but not fastened; and on being completed the parts were taken asunder, numbered, and put aside, to be raised to their places on the completion of the whole. This was necessary, as Colonel Emy found himself unable to raise the principal entire.

For erecting the principals a movable scaffold was prepared, and provided with a template similar to the one on which the arc was formed. This is shown in figs. 3 and 4.

When the scaffold was brought exactly to the place where the principal was to be erected, all the pieces of the latter were raised as numbered, and put in their places; and then, when completely fastened together, the template or centre was detached, and the arc allowed to rest on its wall-plates. The bolts were then added to the arc before it received the weight of the roof timbers. Colonel Emy considered, after completing his work in the way described, that he might have constructed the arcs and fitted all the pieces on the vertical template at once, and thus have saved time and made the work more perfect; but the idea came too late.

The principal being thus placed, was maintained in its position by wedges, by the ends of the radials in the recesses in the walls before mentioned, and by stays nailed to the principals temporarily; the scaffold was then removed to the place of the next principal.

Figs. 21 to 24, Plate XXXIV, are parts of the roof of the riding-house at Libourne. This roof, on the same principle, was erected in 1826; but as the walls at Libourne were of great thickness, and strengthened by counterforts, it was not necessary so carefully to guard against lateral thrust, and therefore the arcs were composed of fewer plates than those at Marac.

The diameter of the intrados of the arcs is 68 feet 8 inches, and the principals are placed 10 feet 6 inches apart, from centre to centre.

In constructing these principals, a working floor was erected at about 2 inches below the wall-plate, and the draught was laid down on it, and a polygonal mould was formed as shown in fig. 771. This is not so perfect as the continuous template, such as was used at Marac, and is more apt to rupture the wood at the points of contact, unless it is freshly cut and very flexible.

The floor on which the principals were constructed extended only half the length of the building; therefore, although the principals were easily raised to the vertical position, by the application of shears and windlasses, each had to be moved to its proper place. This was

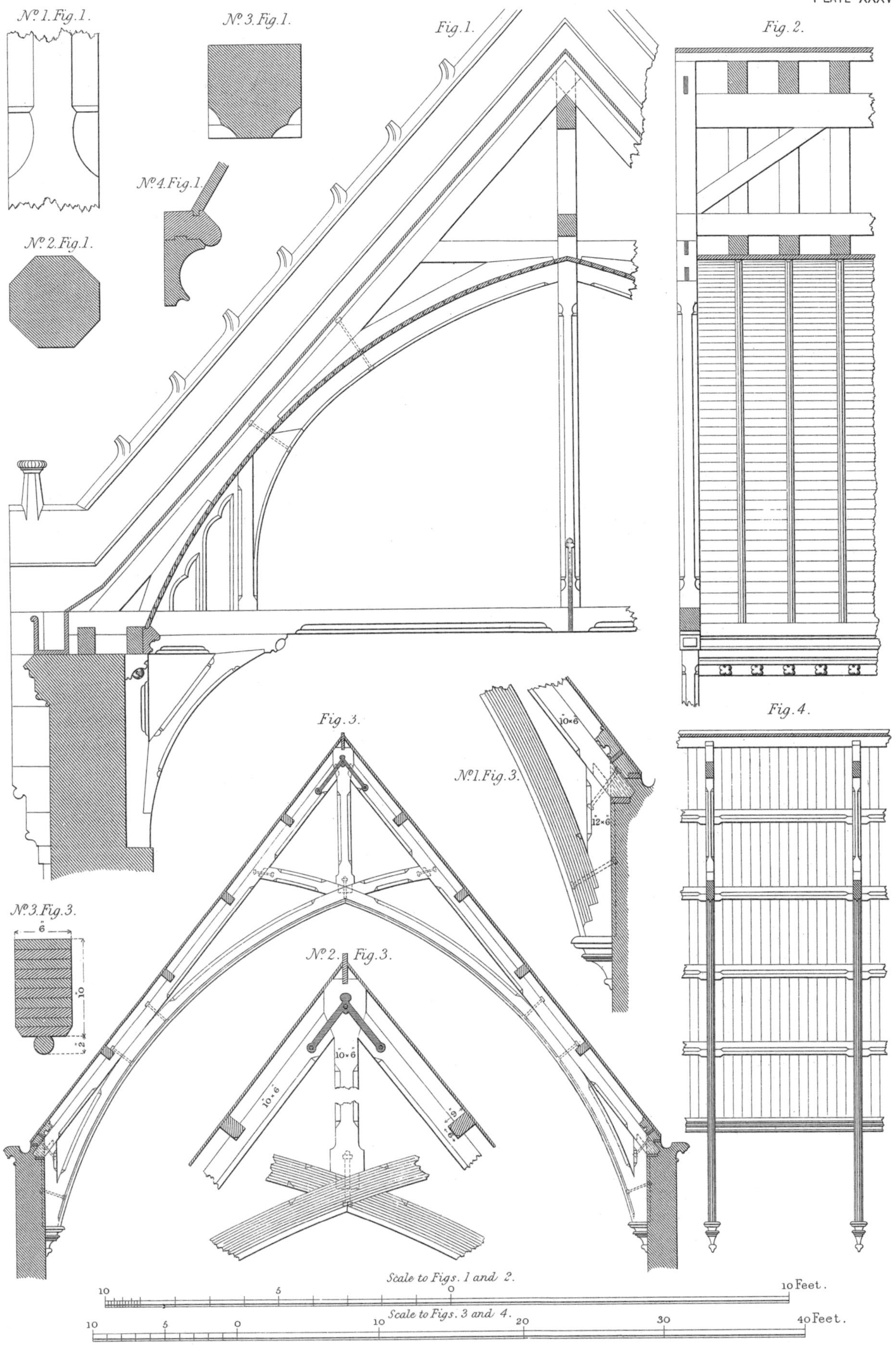

ROOF OF THE SALLE DES CATECHISMES, CATHEDRAL OF AMIENS Figs. 1, 2, and Details

ROOF OF GRASSENDALE CHURCH Figs. 3, 4, and Details

effected by placing under each springing a little carriage running on the wall-plates. The frames were kept upright by proper stays, and the carriages dragged along by tackling, and forced by levers till they reached their places.

Figs. 21 and 22.—Front and side elevations of the springing of one of the principals. The centre of the arc is on the line A B. The principal is mounted on one of the carriages mentioned above.

Fig. 23.—Section through the principal on the line C D.

Fig. 24.—Horizontal section on the line A B.

PLATE XXXV.—Figs. 1 and 2.—Roof of the Salle des Catechismes, Amiens Cathedral. Fig. 1 is a transverse, and fig. 2 a longitudinal section of this roof. It is composed of principals formed with principal rafters, curved ribs, king-post, tie-beam, and collar-beam, and is merely a variety of king-post truss. The ends of the tie-beam, in addition to their wall-hold, are supported by framed brackets resting on stone corbels in the wall. The brackets, tie-beam, king-post, and curved ribs are all exposed to view, and are moulded in a

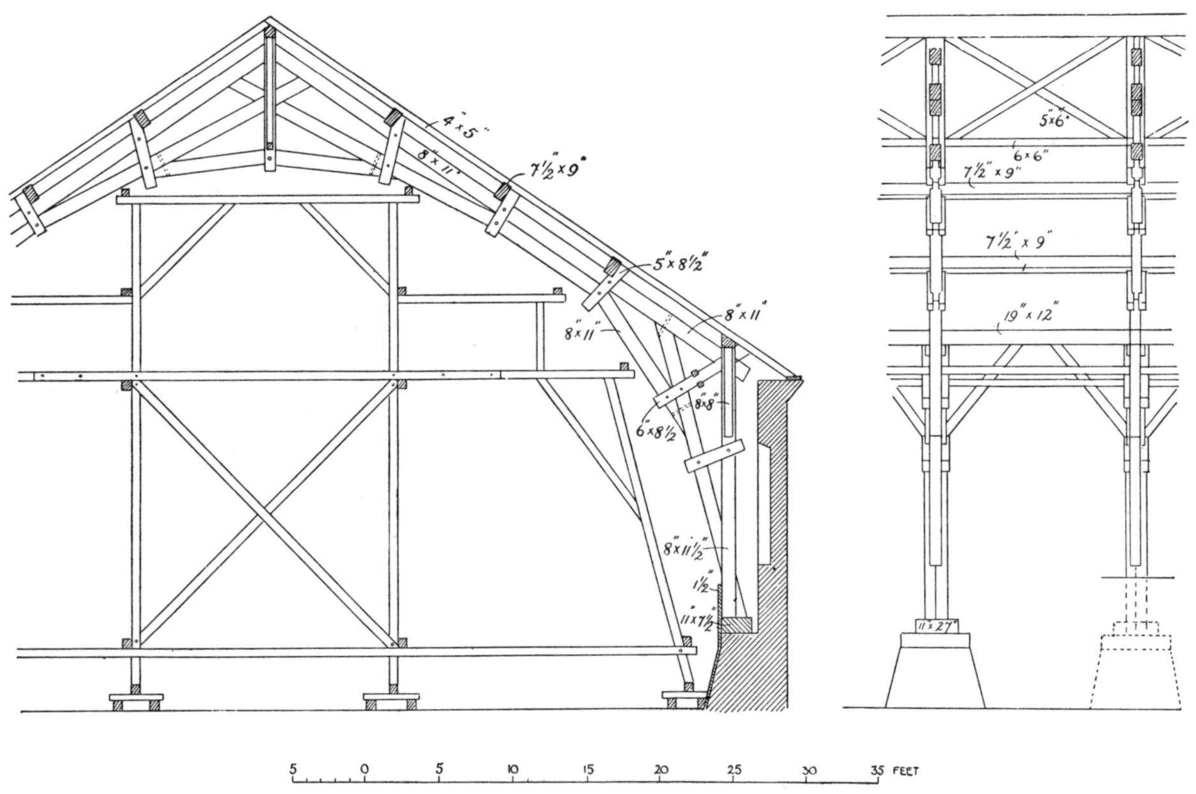

Fig. 772.—Roof of Military Riding-school in Jutland

very simple and effective manner. The ceiling is vaulted, and formed of boarding, ornamented with moulded ribs (in vertical planes) placed about 18 inches apart.

Fig. 1.—No. 1 is an elevation of the lower portion of the king-post, showing the mode of finishing the chamfering. No. 2 is a section through the octagonal portion of the post. No. 3 is a vertical section of the tie-beam. No. 4 is a section of the cornice from which the arched ceiling springs.

Fig. 3 is the elevation, and fig. 4 the longitudinal section of a Gothic roof on Colonel Emy's principle, designed by Mr. Arthur Hill Holme of Liverpool, and erected under his direction at Grassendale Church, Aigburth; and Nos. 1, 2, and 3, fig. 3, are the details of the same roof drawn to a larger scale.

Plate XXXVI shows another variant of Colonel Emy's system, designed by Mr. James Miller, Architect, Glasgow. It represents the roof of the grand avenue of the Glasgow

International Exhibition of 1901. The scantlings throughout were very light, but proved sufficient for the temporary purpose for which the building was erected. The roof was covered in part with glazed top-lights and in part with corrugated iron, and the walls were plastered outside.

The roof of a military riding-school in Jutland (fig. 772) is on somewhat similar lines but with solid timbers of heavier scantling, and is more suitable for a permanent structure. The "arch" is composed of a series of short straight timbers, and is a semi-polygon. The radial braces are composed of two flitches, notched on to the other timbers and secured with bolts. The clear span is about 64 feet 6 inches, and the trusses are 14 feet from centre to centre. Lateral movement of the trusses is prevented by the pole-plate, purlins, and ridge-piece, and by the bracing under the last. The scaffolding used in erecting the trusses is shown in the transverse section. It will be observed that the trusses stand clear of the walls, so that no damage will be done by the slight settlement which is certain to occur with this form of truss. This example is taken from the *British Architect* for March 1, 1878.

7. CONICAL ROOFS, DOMES, TOWERS, AND SPIRES

Conical Roofs.—Small conical roofs may be constructed with common rafters spiked to a circular wall-plate at the foot and to a central post at the apex. Two or more pairs of opposite rafters may be braced by horizontal collars nailed to them at one-third or one-half the height. The example given in fig. 773 was erected from the writer's design to mask the acute angle of an irregular building. In this case the central post is supported on the back of the hip-rafter, and is 4 inches square. The eaves-trough projects 18 inches, and the lower portion of the roof is formed with sprocket-pieces packed up to the required curve. The boards for the projecting eaves and curved portion

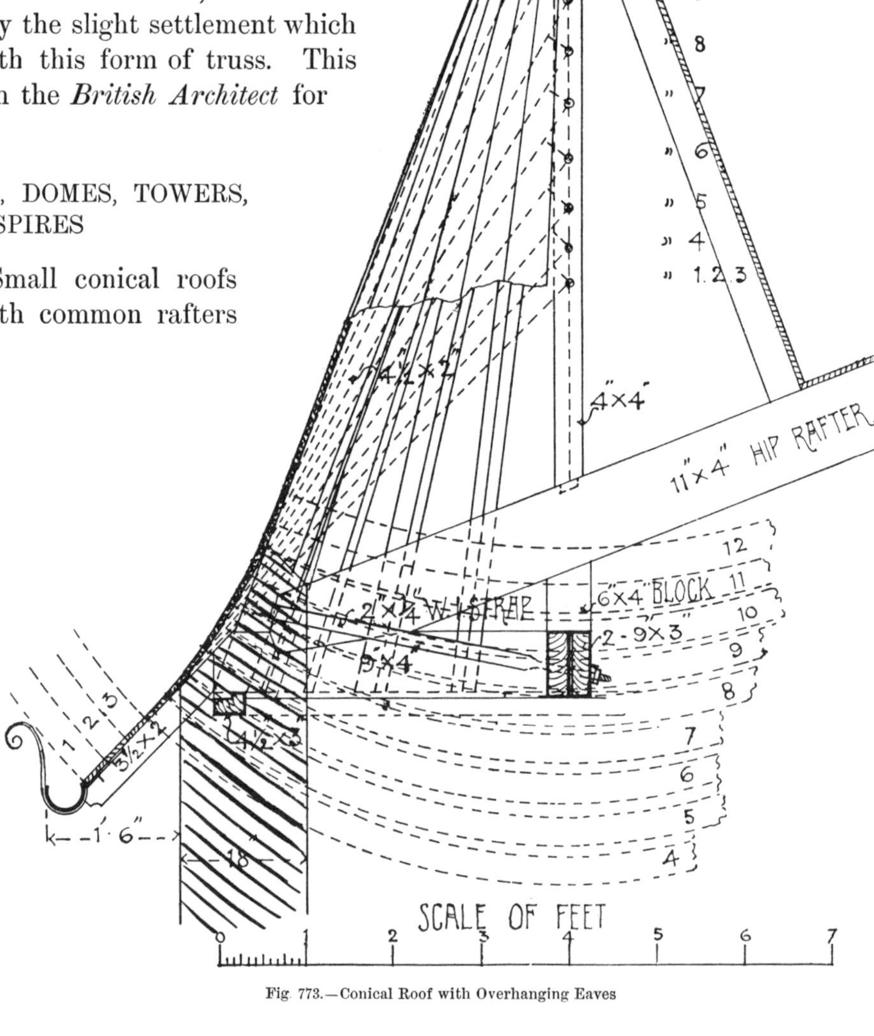

Fig. 773.—Conical Roof with Overhanging Eaves

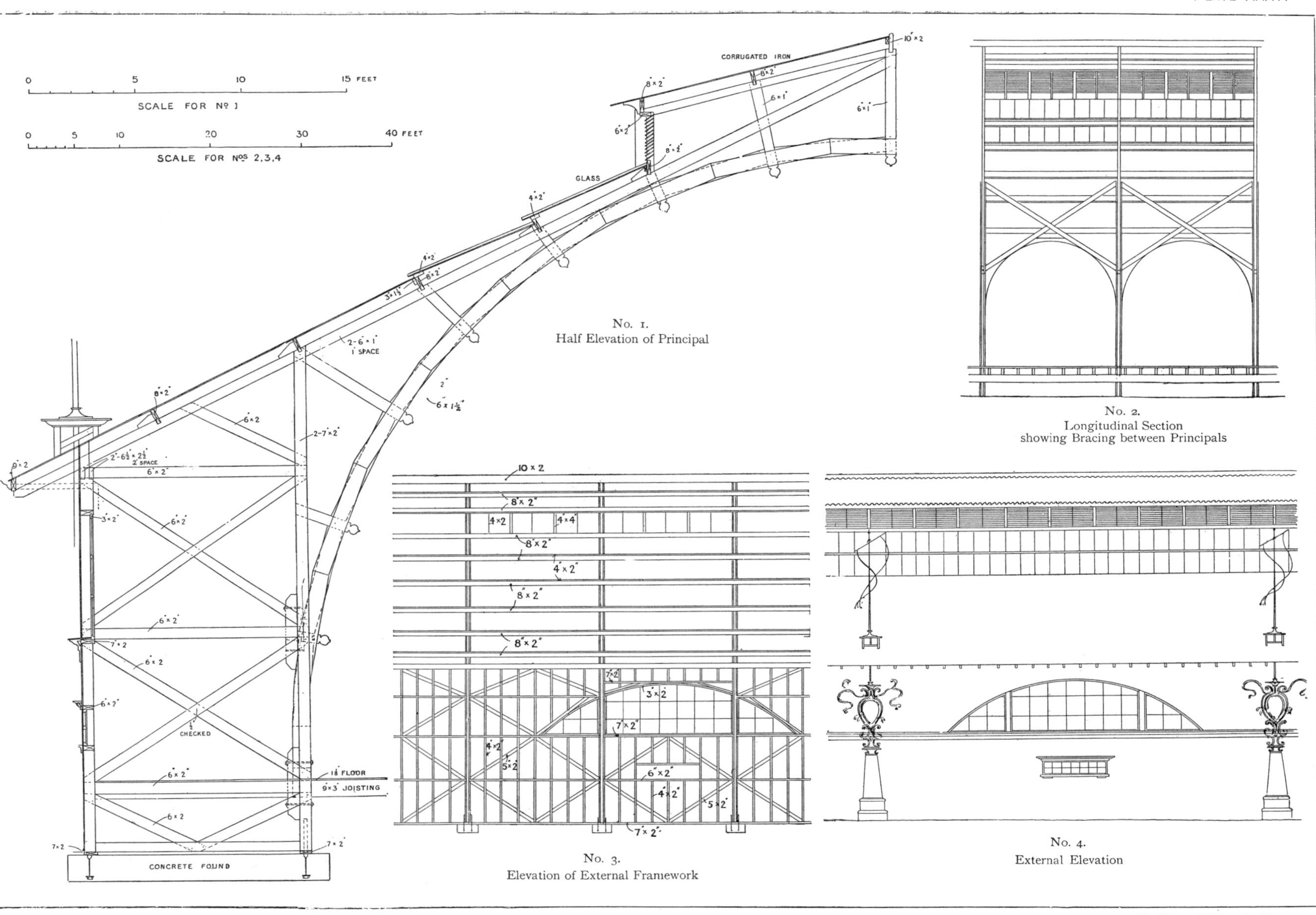

No. 1.
Half Elevation of Principal

No. 2.
Longitudinal Section
showing Bracing between Principals

No. 3.
Elevation of External Framework

No. 4.
External Elevation

Mr. James Miller, Architect

ROOF OF GRAND AVENUE, GLASGOW INTERNATIONAL EXHIBITION, 1901

of the roof are concentric with the eaves-trough; each board is an arc of a circle, the centre being found by producing the surface of the board till it cuts the vertical axis of the cone, as shown by the dotted lines in the section. The boards are in short lengths, and saw-cuts are made transversely to facilitate bending around the rafters. In the upper part of the roof the boards are cut to a gusset shape and nailed to pieces fixed transversely between the rafters. The boarding is covered with felt and slates, the latter being nailed through the felt to the boards.

The next example (fig. 774) is the roof of the steel water-tower at Laredo, Texas, and was designed by Mr. Edward Flad. The tower is 120 feet high to the steel angle shown in the section, and the diameter at this level is 20 feet. The roof is twelve-sided, each angle having a hip-rafter and tie framed into a central polygonal post 12 inches in diameter. Packings are nailed to the backs of the rafters to form the curved outline. The roof is covered with 1½-inch boards and "tin-plate".

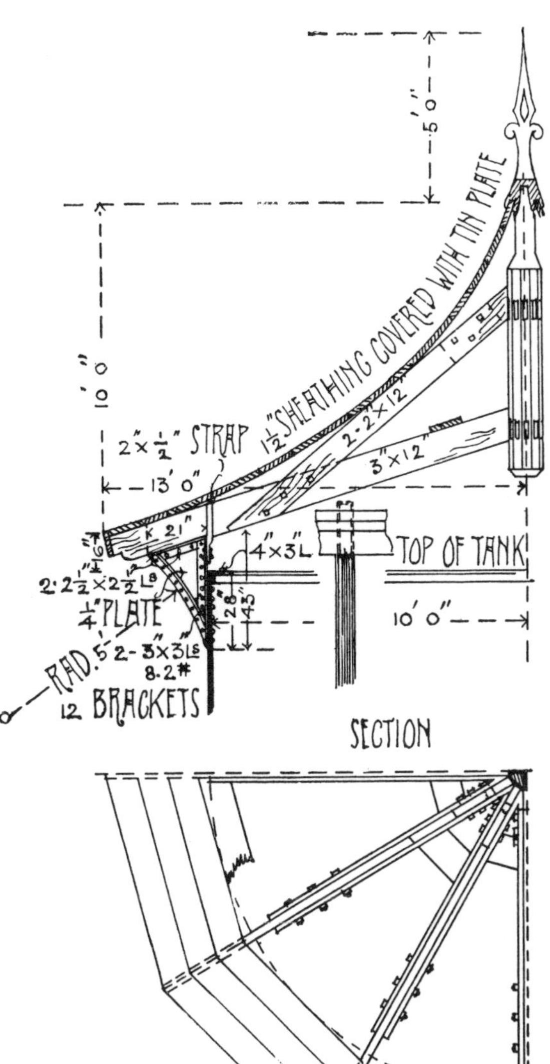

Fig. 774.—Roof of Water-tower, Laredo, Texas

The construction of the conical roof, fig. 775, will be evident from the drawings without detailed description. The two main principals (No. 3) are united at the top by being inserted into iron sockets cast in one piece; and the frame is completed by struts and an iron tie-rod. The other two principals are framed like a queen-post roof, as shown in No. 4; and the ties of all the principals are connected at the centre by the radiating straps seen in No. 5, through the central circular part of which the tie-rod No. 6 passes, and is secured by a nut. The same letters refer to the same parts in most of the figures, excepting *a*, which, in Nos. 1 and 2, is the larger circular purlin, but in No. 4 the straining-sill between the queen-posts.

If a hemisphere or any other portion of a sphere *acb* (fig. 776, No. 1) be intersected by vertical planes *e*, *d*, equidistant from its centre, the angular portions *h h*, between the boundaries of the planes, are pendentives.

In like manner, in a conoid *a c b* (fig. 776, No. 2), the angular portions *fff*, between the intersecting planes *e*, *d*, are pendentives, and the same in an ellipsoid. In these figures, the convex surfaces of the hemisphere and conoid are shown, but in vaulting it is, of course, the concave surfaces which form the pendentives, as in fig. 777, where A and B are two of the contiguous intersecting planes; C, part of the concave surface of the vault; and D, one of the pendentives. The resulting curve of the intersection of a spherical vault by a plane is a portion of a circle, that of an ellipsoid is an ellipse when the plane is parallel to the major axis, and that of a conoid a hyperbola.

To describe a spherical dome over a square room, with spherical pendentives and a circular skylight in centre.

Let A B C D (fig. 778, No. 1) be the plan of a room 15 feet square: draw the diagonals of the square, and from their intersection E describe the inscribing circle A *k* B C D, which will be the plan of the hemispherical vault. On any of the sides of the square A B, B C, describe

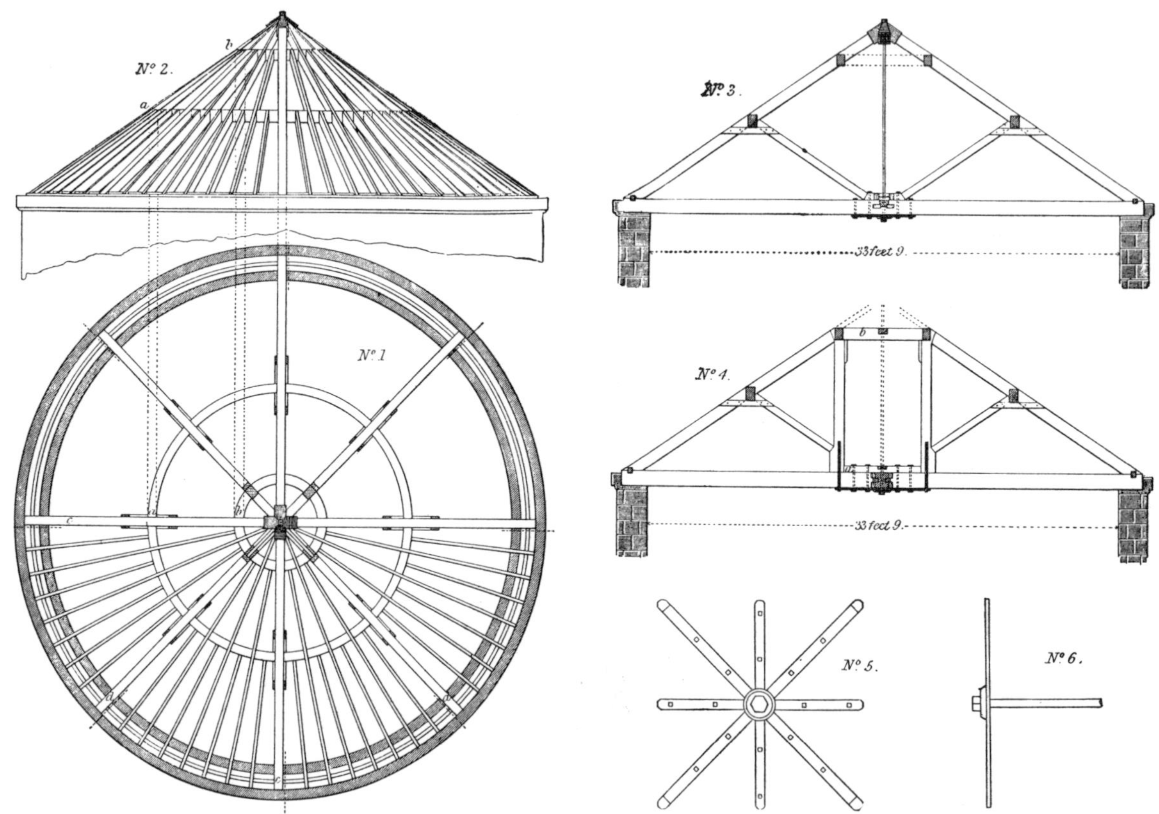

Fig. 775.—Conical Roof with King-post and Queen-post Principals

a semicircle which will be the curve resulting from the intersection of the hemisphere by the plane of the side of the square. To find the plan of the ribs: From the centre E describe a circle of the size required for the skylight, and draw the double line showing the breadth of the curb *b*. Divide the circle into as many equal parts as there are ribs required, and from

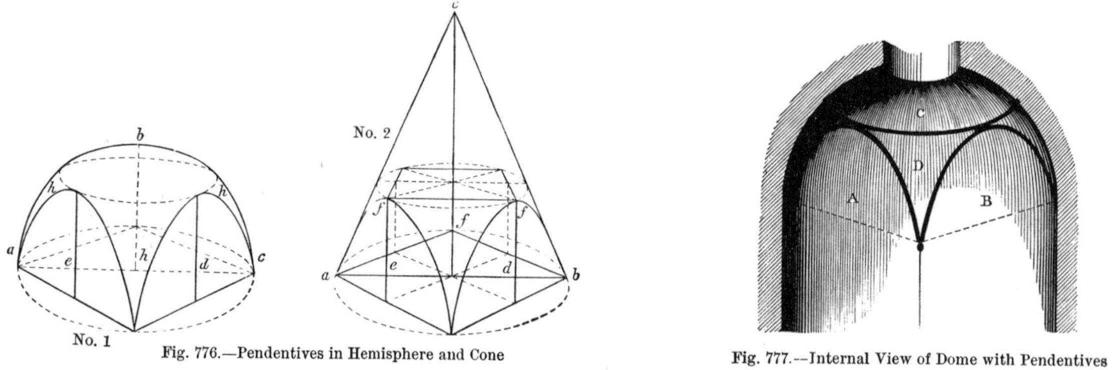

Fig. 776.—Pendentives in Hemisphere and Cone

Fig. 777.—Internal View of Dome with Pendentives

the points of division draw radii for the centres of the ribs. Set off half the thickness of the rib on each side of the radial lines, and draw parallel lines representing the sides of the rib. If the ceiling is to be finished with plaster, the ribs should be nowhere more than 14 inches apart.

No. 2 is a section on the line H I of the plan. Bisect the line F G in *s*, and draw the semicircle for the resulting line of the intersection of the hemisphere. From *s*, with the radius E A, describe the segment *a a a* representing the section of the spherical surface on

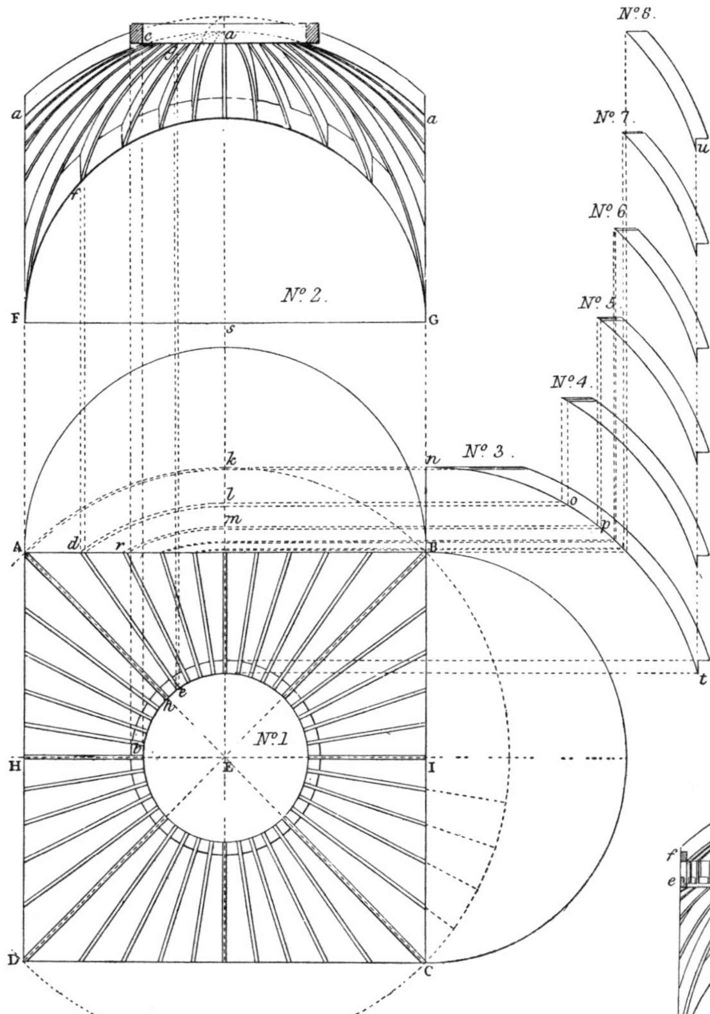

Fig. 778.—Spherical Dome with Pendentives and Skylight, on Square Plan

the line H I; draw the curb *c* from the plan, as shown by the dotted lines; and find the intersection of the other ribs with the side arch and the curb, by drawing lines from the plan, as from *d* to *f* and from *e* to *g*. The projections of all the ribs except the central rib and side ribs *a a* will be elliptical curves.

To find the length of each rib.—From the centre E, with the radius E A, describe the arc A *k*, and draw *k n* parallel to the side of the plan. Then with the same radius, from I as a centre, describe the arc *nt* for the under side of the rib A*h*. From E describe the arcs *dl*, *rm*, &c., and draw *l o*, *m p*, &c., to intersect *n t* in *o*, *p*, &c.; then *o t* and *pt* will be the lengths of the ribs *d e*, *r f* respectively. By drawing the lines *o* 4, *p* 5, &c., and *t u*, and describing arcs with the radius E A, all the ribs may be drawn separately as Nos. 4, 5, 6, 7, 8. The double dotted curves A *k*, *d l*, *r m*, show how the bevel of the end of the ribs is obtained.

Fig. 779 is a spherical vault less than a semicircle, with a plain fascia introduced in the vault above the pendentives: the curve resulting from the intersecting planes is the segment of a circle. A B C D is the plan of the room 16 feet square, F the centre of the segment, G *m* H L the plan of the double curb over the pendentives, and *k* the curb of the skylight. From the centre E on the line E A, make E F equal to *r* F, and from F describe the curves D *c*, *d b* of the

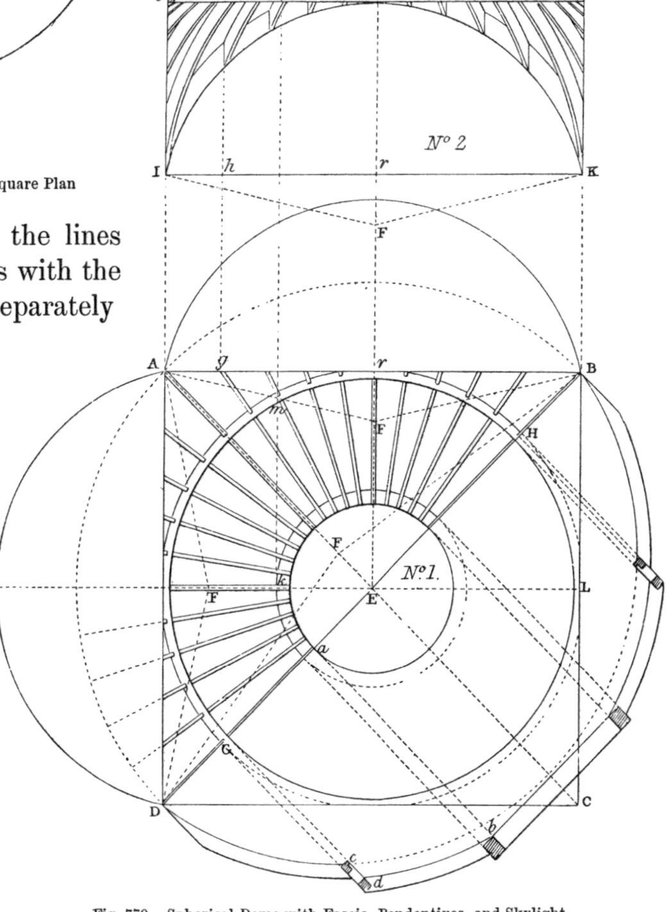

Fig. 779.—Spherical Dome with Fascia, Pendentives, and Skylight

ribs D G, G *a*. The dotted lines drawn from the plan show the manner of obtaining the section of the curbs *c d* and *b*. In the section No. 2, make *r* F equal to *r* F in the plan, and F is the centre of the segments I K, *e l o*. The position of the ribs is found by drawing lines from the plan, as *g h*, *m p*.

To describe a conical dome over a square room, with a circular skylight in the centre.

Fig. 780 shows the plan and section of a conical pendentive: A B C D is the plan of the room 16 feet square, and the base of the cone is the inscribing circle ABCD, described from E, and shown by a dotted line. To find the ribs:—Make E O, No. 1, equal to the height of the cone, and join A O, C O. From the plan of the curb draw the dotted lines intersecting A O and C O in F and G. From the centre E describe the arcs *b l*, *c k*, &c., meeting E C in *l k i h g*; and draw the perpendiculars *l m*, *k n*, &c., cutting the line C G. Then *m* G will be the length of the rib *b*, *n* G the length of *c*, and so on.

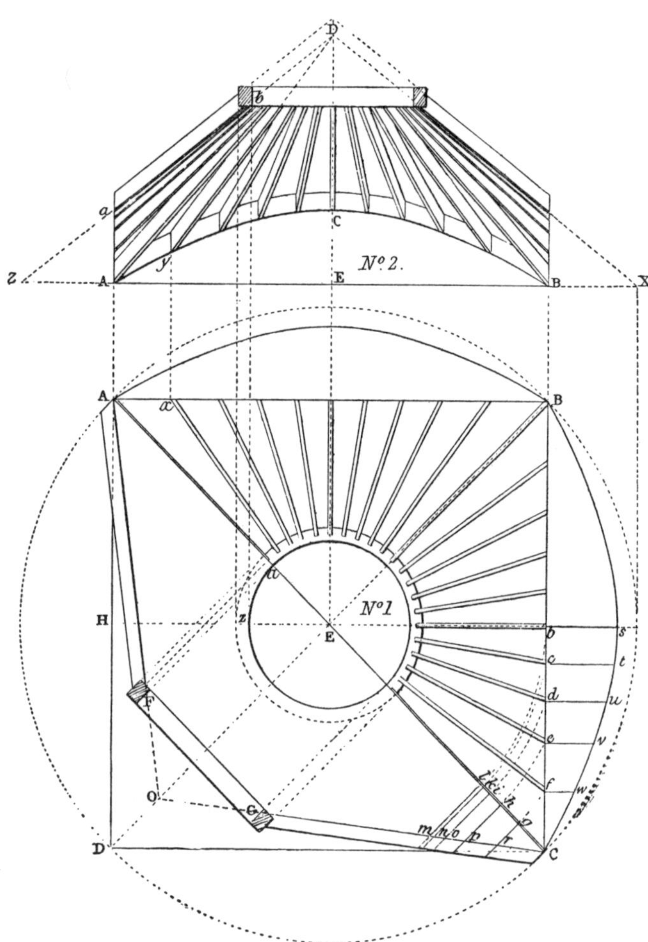

Fig. 780.—Conical Dome with Pendentives and Skylights

To find the hyperbola resulting from the intersection of the sides:—Make the perpendicular *b s* equal to *l m*, *c t* equal to *k n*, *d u* to *i o*, *e v* to *h p*, and *f w* to *g r*; then trace the curve through *s t u v w*.

In the section No. 2, make E X, E Z equal to half the diameter of the base of the cone, and make E D equal to E O, the height of the cone; join Z D and X D. Find the hyperbola A C B, as in the plan, or transfer it by ordinates. Find also the intersections of the ribs with the curve A C B, by drawing lines from the plan, as *x y*; and find their intersections with the curb *b* in the same manner. Then from the intersections of the ribs with the line A C B, as at *y*, draw lines converging to D, for the vertical projection of the ribs.

To describe the pendentives formed by the intersection of an octagonal domical vault by a square.

Let ABCD (fig. 781, No. 1) be the plan of the apartment 20 feet square, C E F G H I K D the plan of the octagonal vault, and A K I C the curve of one of the longest or diagonal ribs A L C. Then to find any of the angle ribs, as D P:—Produce C *d* D to M, and divide the portion M K of the rib A C into any number of equal parts, as 1, 2; and through the points of division draw 1 *l*, 2 *m*, perpendicular to A L and cutting D P in *l* and *m*. On D P erect the perpendiculars *l* 1, *m* 2, P *n*, and make them equal to the corresponding ordinates *l* 1, *m* 2, R *n*; and through D 1 2 *n* draw the curve of the rib. The intermediate parallel ribs are all portions of the same curve, and their lengths and bevels are found by drawing lines *e* 1, *f* 2, *g* 3, &c., from the intersections of the lines of the ribs with the side of the square to the points 1 2 3 &c., of the rib A C.

To find the projections of the ribs in the section No. 2:—From the points *e f g* C in the plan, draw the perpendiculars *e* 1, *f* 2, *g* 3, C O, and transfer to them the heights of the

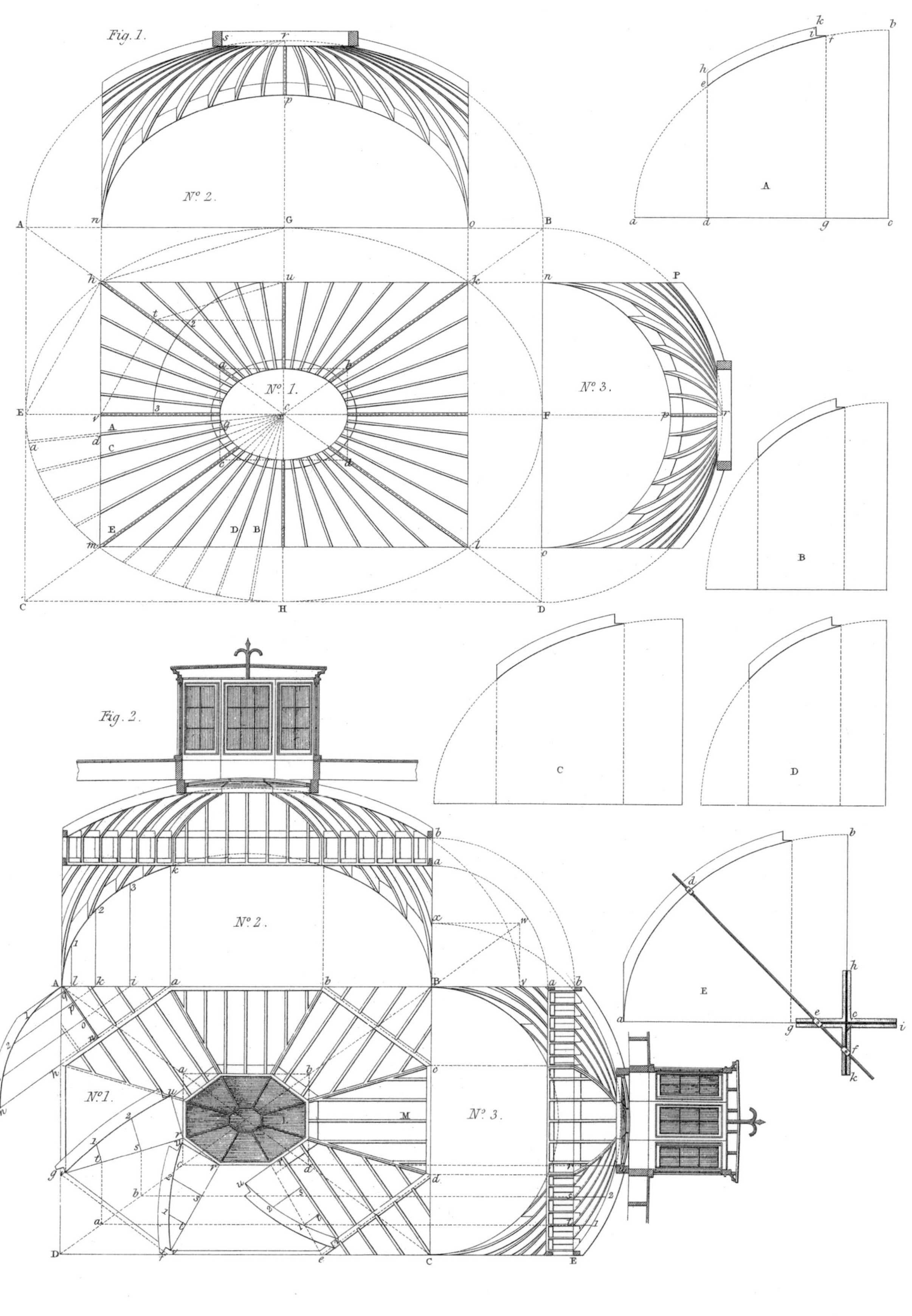

PENDENTIVES

perpendiculars a 1, b 2, c 3, d M, which will give the points 1 2 3 O, through which the curve of the wall rib A O is to be drawn, and the same points also give the intersections of the jack-ribs with the wall rib.

To draw an elliptical domical pendentive roof.

PLATE XXXVII.—Let $h k l m$ (fig. 1, No. 1) be the plan of the apartment: draw the diagonals $h l$, $k m$, and through c the lines E F, G H, parallel to the sides of the apartment. To find the circumscribing elliptical base of the dome:—From the centre c describe the quadrant u 2 3, and bisect it at 2; and through 2 draw 1 2 t parallel to the side of the rectangle $h k$; join $t u$, $t v$. Then from h draw h E, h G parallel to $t v$, $t u$ respectively, cutting the lines E F, G H in E and G; and complete the parallelogram A B D C, which will be the circumscribing rectangle of the ellipse forming the base of the dome. The ellipse of the curb is proportioned by drawing $a b c d$ to meet the diagonals of the rectangle $h k l m$.

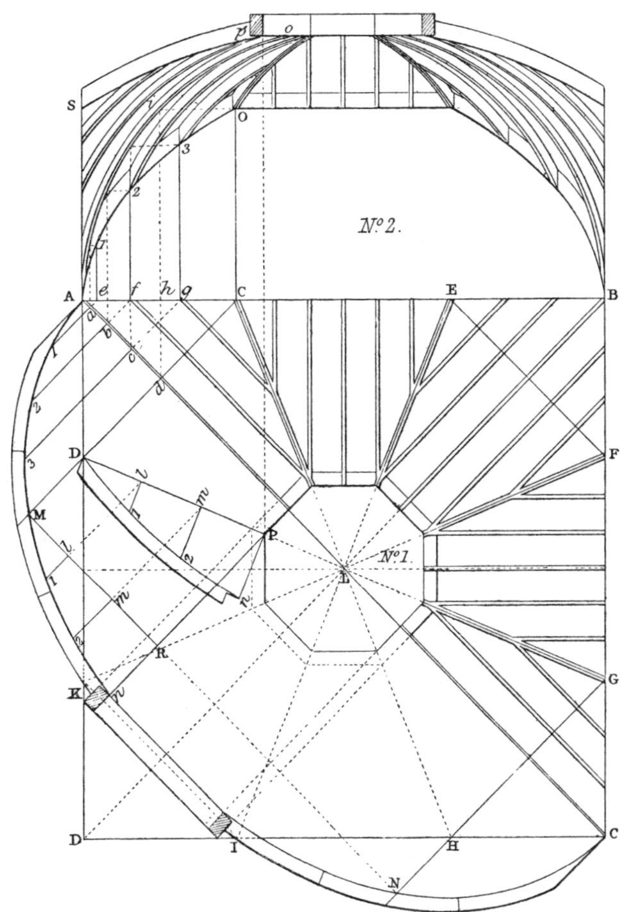

Fig. 781.—Octagonal Domical Vault with Pendentives and Skylight

On the line B D (No. 3) describe the semicircle B r D, as the section of the dome on its minor axis. Then, as the dome may be considered a solid, formed by the revolution of the semi-ellipse E G F round its axis E F, it follows that all sections of it by planes parallel to G H will be circles. Therefore, to find the wall rib $k l$:—On B D, from the centre F and with the radius $c u$, describe the semicircle $n p o$. A r B, No. 2, gives the section on E F, and is a semi-ellipse, and all sections made by planes parallel to E F will be similar figures. Hence $n p o$ (No. 2) represents the wall curbs for the sides $h k$, $m l$, the semi-axis minor G p being equal to $n p'$ in No. 3.

The other ribs may be found by the method of ordinates; but they are much more accurately and easily drawn by the trammel, in the manner shown at E. The points $g h i k$ of the trammel being set on the major and minor axes of the ellipse, and $d e$ being made equal to the minor, and $d f$ to the major semi-axis, the distance between d and e will necessarily remain constant in describing all the figures A B C D E, but f will be moved nearer to e for the curve of each rib from A to B. Fig. A shows the manner of finding the length of the rib marked $a d g$ on the plan. The line $a c$ (fig. A) is made equal to $a c$ on the plan, and is similarly divided at d and g. Perpendiculars $d e h$ and $g f$, drawn from d and g, give the proper lengths of the rib on the curve. The lengths of the ribs B C D E are found in the same way, as shown by the figures B, C, D, and E.

To describe a domical pendentive of an irregular octagonal plan over an apartment, the plan of which is a parallelogram.

A B C D (Plate XXXVII, fig. 2, No. 1) is the plan of the apartment: the mode of proportioning the ellipse of the base of the dome, and the octagons of the curbs, is the same as in the last figure. In No. 2 the elliptic curves of the centre rib on the line of the major axis, and also the wall rib on the side A B, are shown. No. 3 is the section across the centre of the plan. The method of finding the other ribs will be found by inspection.

Domes in carpentry are composed of a number of ribs, placed vertically in planes, which in spherical domes would, if prolonged, pass through the vertical axis of the dome. When the surface of the dome is a surface of revolution, all the ribs have the same exterior contour or profile. In domes on polygonal plans, the angle ribs at the intersections of the sides of the solid, alone, are in planes which pass through the axis. The ribs generally spring from a wall-plate or curb, forming a ring laid on the walls which support the dome, and this ring should be made sufficiently strong to resist the lateral thrust of the ribs, so that the walls may have to support only the downward pressure or weight of the dome. The central point in the curved surface of a dome is called its pole or centre; the imaginary straight line drawn from the pole to the centre of the base is its axis. When the height of a dome is greater than the radius of its base, it is said to be surmounted; when less, surbased. An aperture at the pole of a dome is called its eye.

To draw an oblong surbased dome on a rectangular plan.

PLATE XXXVIII.—This example may be regarded as a groined vault rather than a dome. The plan, fig. 1, No. 1, shows the mode of placing the ribs. No. 2 is the section across the shorter axis of the plan; and No. 3, the section on its longer axis. The curve of the rib E *n e n* F, and of all the ribs parallel to it, is found by dividing the segment C *n e n* D, No. 2, into a number of equal parts, and drawing lines from the divisions to meet the diagonal A B, No. 1, in the points *f g h i*, and from those points drawing lines parallel to C D, cutting E F, and produced indefinitely; then transferring the heights of the ordinates on C D to the corresponding ordinates on E F. The curve of the diagonal rib A *k* B is also found in the manner above described, as the lines show. The positions of the purlins *n*, *m*, and their projections on the plan, are found by drawing lines from their sections at *n n* to meet the diagonal.

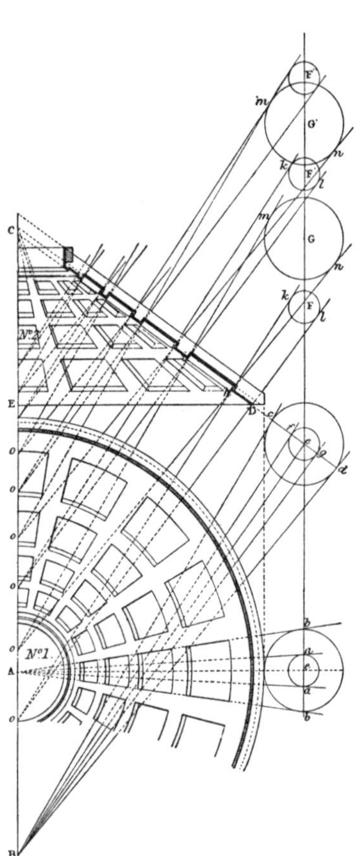

Fig. 782.—Panelled Conical Dome

To draw a surbased dome on an octagonal plan.

In fig. 2, No. 1, the position of the ribs is shown, and also the manner of finding the curve of the angle ribs. No. 2 is a section on the line A B. On the plan No. 1, the rib standing over C G is drawn at G 1 2 3 4 *c*, and divided into equal parts, from which ordinates are drawn to the chord line, and produced to the line of the angle rib H I; from the points *h i k l*, in which these intersect H I, ordinates are drawn; and the heights F *c*, *d* 4, *e* 3, &c., being transferred to them, give points through which the curve of the angle rib is to be traced.

To describe domes on circular plans.

Nos. 1 and 2, fig. 3, are the plan and section of a hemispherical dome; and Nos. 1 and 2, fig. 4, are the plan and section of a dome which is the half of a prolate spheroid on a circular plan. The construction is obvious.

To divide domes of circular plan into caissons or panels.

Let No. 1, fig. 782, represent part of the plafond or soffit of a conical vault, and No. 2 part of a vertical section through its axis. Having divided the plafond into the number of panels and ribs or fields between them that is suitable to the design, the heights of the caissons and of the horizontal fields between them are found as follows:—On A *e*, No. 1, draw the circle *b b*, making it a tangent to the circular base of the vault, and to the lines A *b*, A *b*; and through its centre draw *b* F″ perpendicular to A *e*. Draw also the circle

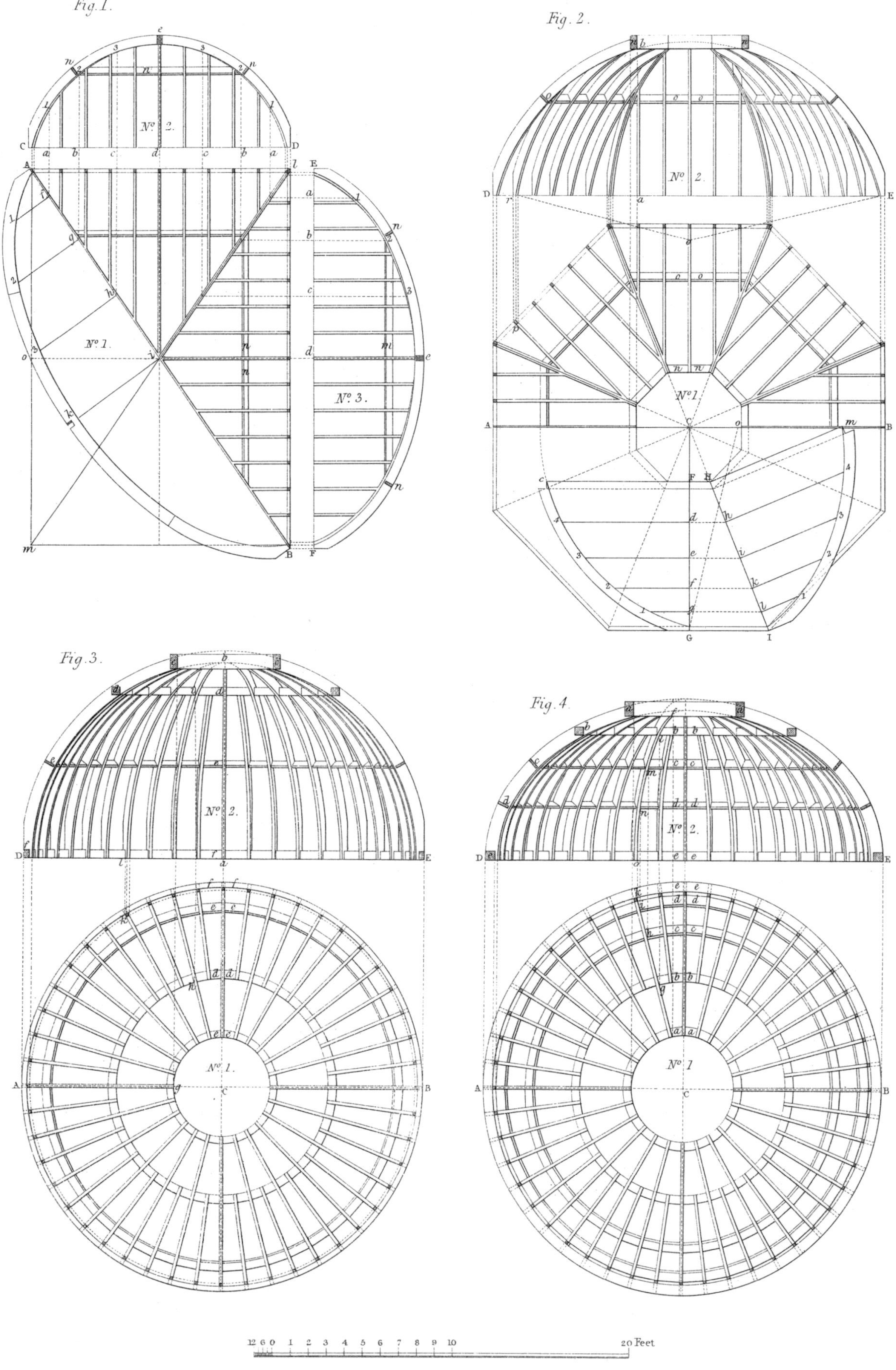

DOMES

a a tangential to the lines A *a*, A *a*, the width between which is equal to the width of the ribs or fields between the panels. Produce the side of the cone C D, No. 2, to cut the perpendicular *b* F″ in *e*, and from *e* as a centre describe two circles, *c d*, *f g*, equal to the circles *b b*, *a a*, No. 1; and through the centre *e* draw the line *e* B perpendicular to the side of the cone C D, and cutting the axis C E produced in B: draw also B *c*, B *f*, B *g*, B *d*. Then, to find the height of the first field: In the section No. 2, through the point D draw a line *o* D *l* parallel to B *g*, cutting the produced axis in *o*; and with the centre F, on the line *b* F″, describe the circle *k l* equal to *f g*; and draw *k o*, cutting the side of the cone in *h*, which is the height of the first field. Through the point *h* last found, draw the line *n h o* parallel to B *d*, and from the centre G, on the line *b* F″, describe the circle *m n* equal to *c d*, and having *n o* for its tangent; and join *m o*: the point where this line cuts the side of the cone is the height of the first caisson. Proceed in the same manner with the circles F′, G′, &c.

To determine the heights of the divisions of a spherical dome.—Fig. 783, No. 1, is part of the plafond of a spherical vault, one-half showing the divisions, and the other the mode of framing. To find the height of the divisions:—Produce the meridian lines *d b*, *d b*, representing the width of the panels, and draw the circle *b c b*, touching the generating circle in G and the lines in *b b*; draw also the meridians *d c*, *d c*, representing the width of the field between the panels, and describe the circle H. Through the centre of the circles, *a*, draw *a b* indefinitely, and perpendicular to *d a*. Then, having fixed the first horizontal division on the profile of the vault, No. 2, draw through it the line E*mn*, and draw the circle H, touching E*mn* in *m*, and the larger circle G, touching it in *n*; H and G being respectively equal to the lesser and greater circle in No. 1. Then draw the second line E *m n* tangential to the circle G, and describe the circle H′ (with the same radius as H), touching this line in *n*; and so proceed, drawing the tangents and the circles H and G alternately; and the intersections of the tangents with the profile of the vault determine the heights of the divisions.

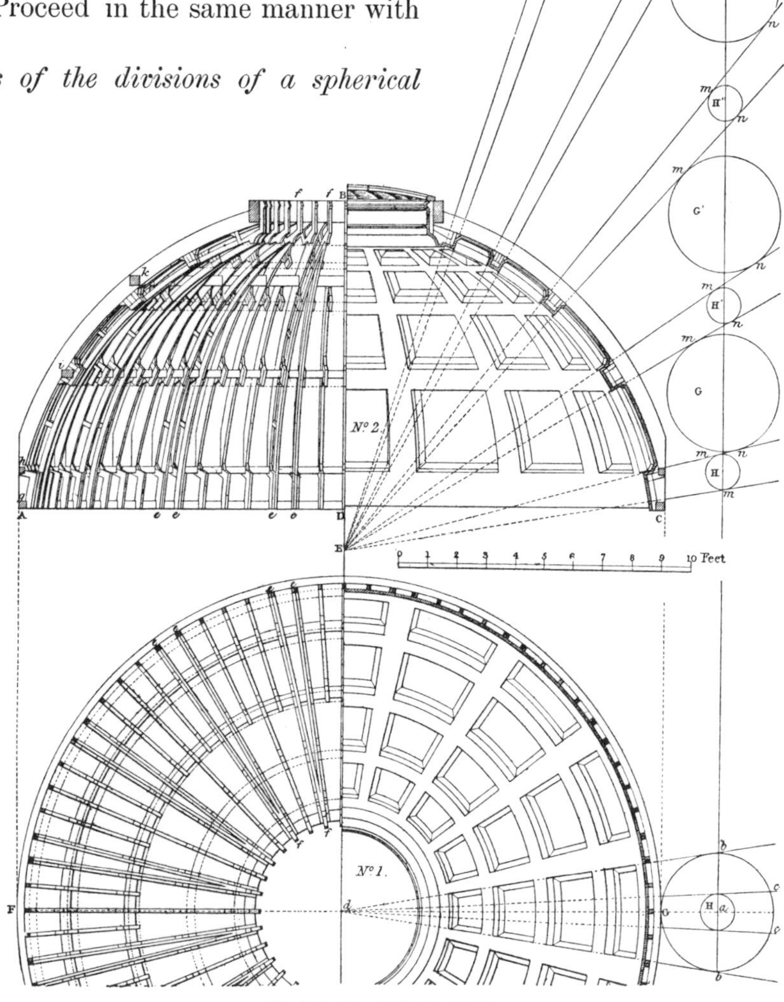

Fig. 783.—Panelled Spherical Dome

To determine the caissons of an ellipsoidal dome.

In fig. 784, No. 1 is the plan of the dome, No. 2 its transverse, and No. 3 its longitudinal vertical section. The divisions on the transverse profile of the dome are obtained by the intersection of the tangents to the circles O, P, &c., as in fig. 783. Then, to find the divisions on the longitudinal section, draw the circumscribing parallelogram R S U T, No. 1, and the diagonals R U, S T. Draw the divisions from the profile, No. 2, to the transverse axis A C, No. 1, cutting the diagonals in *l m n o p r s t*, and through these points draw lines parallel to A C, and the points wherein these intersect the major axis B D give the divisions on the longitudinal profile, No. 3.

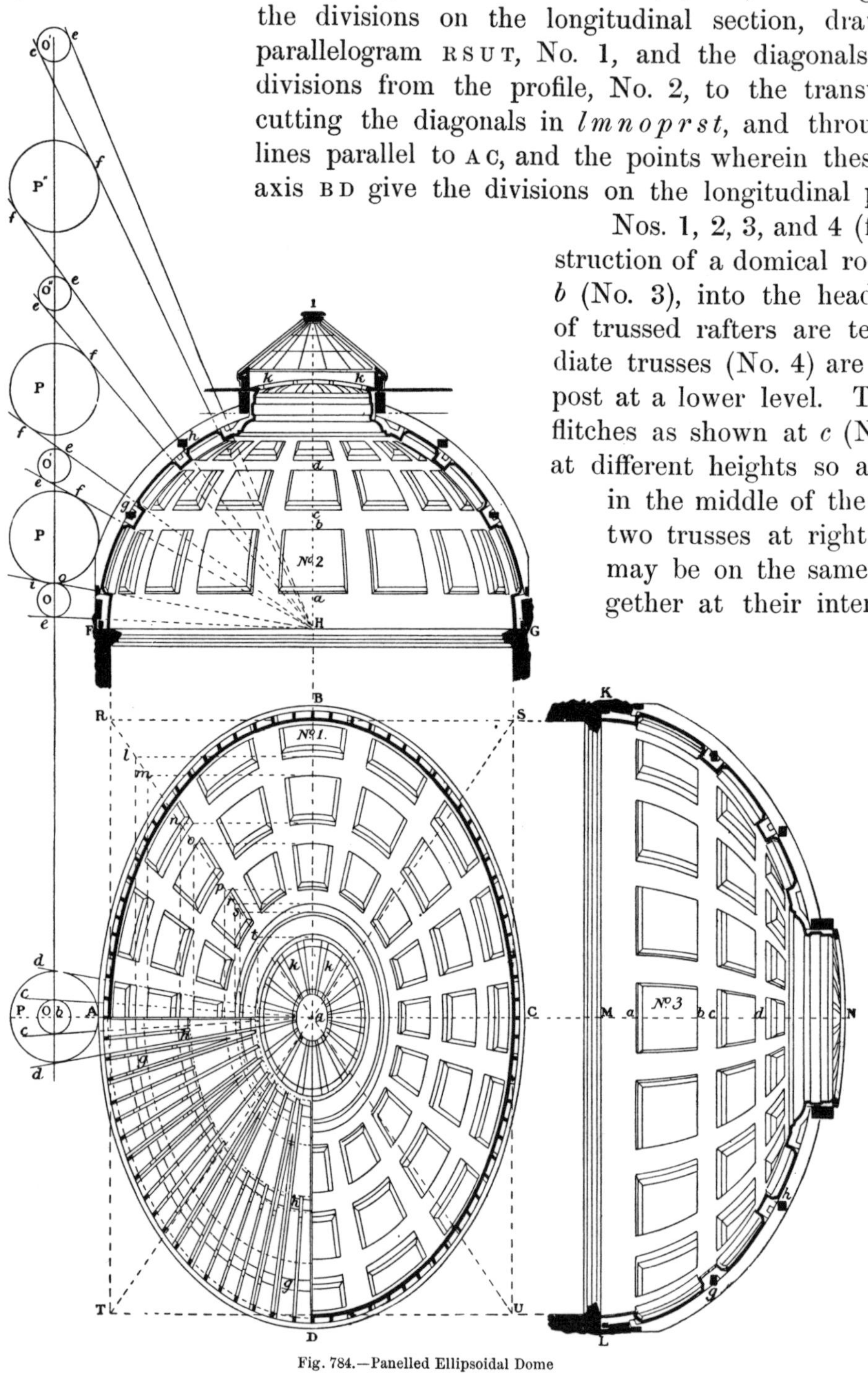

Nos. 1, 2, 3, and 4 (fig. 785) show the construction of a domical roof with a central post *b* (No. 3), into the head of which four pairs of trussed rafters are tenoned; four intermediate trusses (No. 4) are framed into the same post at a lower level. The collars are in two flitches as shown at *c* (No. 3), and are placed at different heights so as to pass each other in the middle of the span. The collars of two trusses at right angles to each other may be on the same level, and halved together at their intersection, as shown at *d* (No. 4). The curved ribs are supported by struts from the principals, as seen in Nos. 3 and 4. The plan and elevation (Nos. 1 and 2) exhibit the curved arrises which the sides of the horizontal ribs assume when cut to the curvature of the dome, as at *a*, No. 3.

Fig. 784.—Panelled Ellipsoidal Dome

PLATE XXXIX, fig. 1.—Nos. 1, 2, 3 and 4 show the construction of a domical roof with a circular opening in the centre for a skylight. Two of the main principals, C D and the corresponding one, are framed with a king-post *c*, as shown in No. 3; the others at right angles to these, with queen-posts, as seen in No. 4. The main ribs correspond to the principals, and the shorter ribs are framed against curbs between them, as at *a*, Nos. 1 and 3.

Fig. 2.—Nos. 1, 2, and 3 show the framing of an ogee domical roof on an octagonal plan. The construction will be readily understood by inspection; and the method of finding the arris ribs, shown in No. 3, will be understood from what has already been said when treating of hip-rafters.

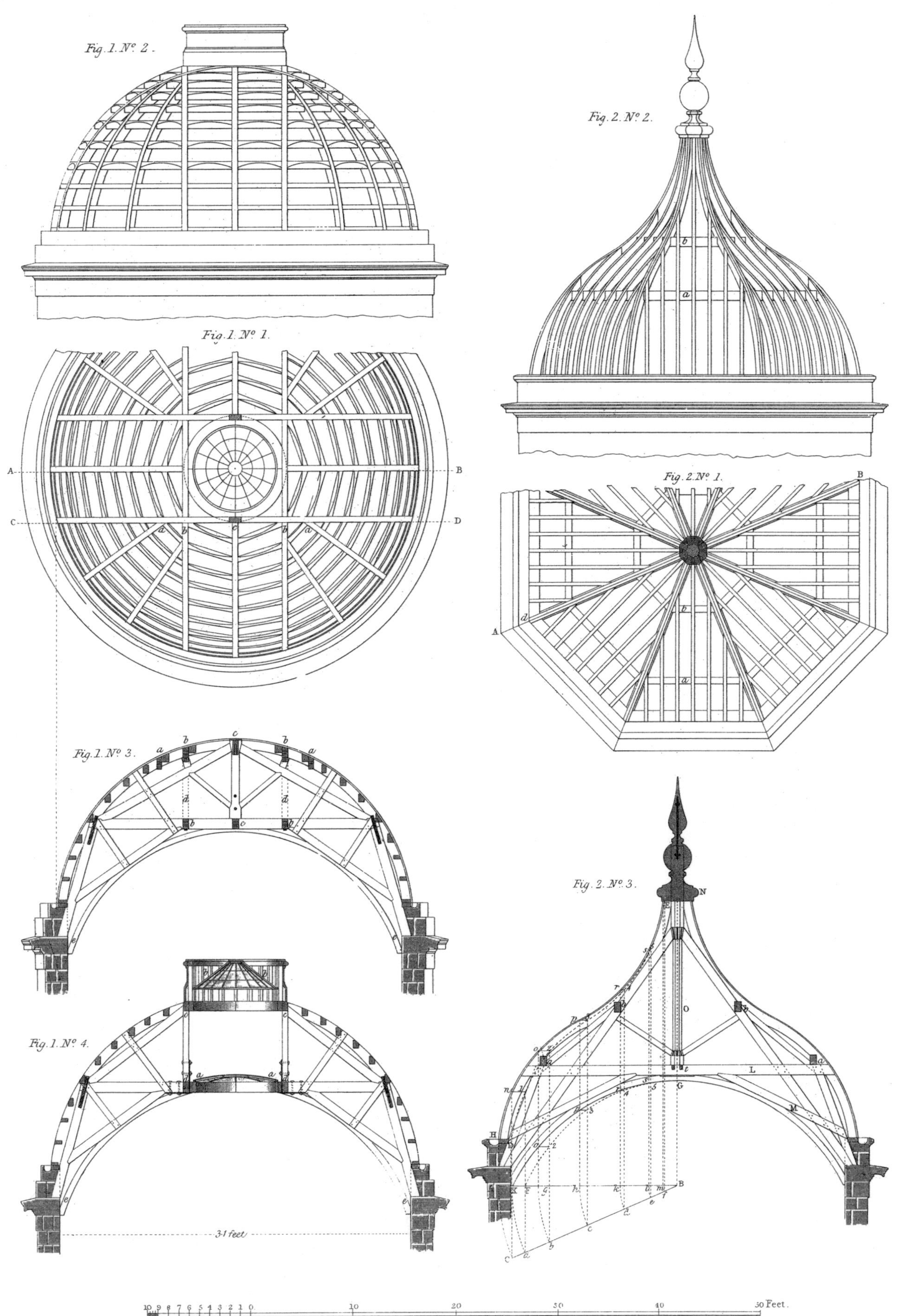

CIRCULAR AND POLYGONAL DOMES

PLATE XL.—The figures on this plate illustrate the construction of a timber steeple.

Fig. 1 is a horizontal section of the square part of the steeple on the line A A in figs. 2 and 6, and also a plan of part of the roof, showing the four principals which carry the steeple. Each principal carries three main posts *a d a* (fig. 1), forming the carcass of the square part, and the two interior principals carry also the additional posts *b b*, to form the octagonal part above.

Fig. 2 is a vertical section of the steeple and roof on a line coinciding with the face of the interior principal *a c d c a* in fig. 1. A A are the principal rafters; B B, the tie-beam;

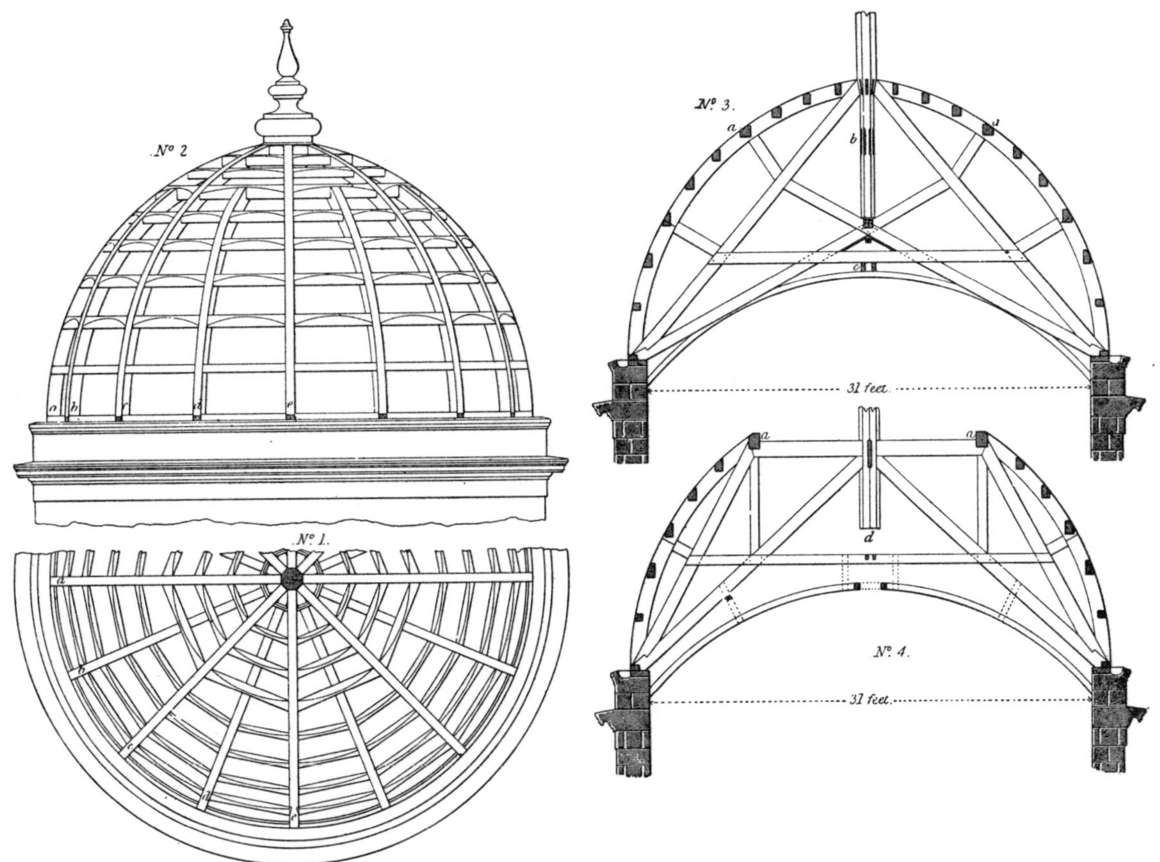

Fig. 785.—Domical Roof and Ceiling on Circular Plan

C C, queen-posts carried up to form the framing of the square portion of the steeple, and marked *a a* in fig. 1; D, king-post, marked *d* in fig. 1; F, secondary rafters; E E, straining-piece; G K H, struts; P, the common rafter; *r*, the purlin; *x*, the gutter-bearer; *y*, the straining-sill; *w u*, a bolt uniting the rafters, tie-beam, and pillow-piece; *g*, a strap uniting the queen-posts, rafter, and straining-piece, and the post L, which is carried up to form the octagonal part, and is marked *b* in fig. 1. The struts H and M serve to discharge the weight of L to the king-post: M is shown on the plan (fig. 1) at *m m*; and the strut *n*, shown in dotted lines (fig. 2), serves the same office in respect of the outer post of the octagon, and is marked *n n* in fig. 1: N is a counter-brace, the intersection of which with M is marked *c* in fig. 1. The strut *o* supports the cross-piece O at the point where the bearers *h h h*, in fig. 1, rest on it. Horizontal pieces unite the summits of the four king-posts, and are marked *l l* in fig. 1 and *g* in fig. 2. Similar pieces, *k k* in fig. 1 and *g* in fig. 2, unite the summits of the queen-posts, and also the posts forming the octagon at successive stages. The exterior posts of the octagon are supported by the horizontal bearers shown in dotted lines on fig. 1, and marked *e e e e*, *f f*. The posts stand on the small parallelogram *g g g g*. At the height of the head-piece Z, the square portion of the steeple terminates, and the octagonal framing is alone continued. Fig. 3 is a horizontal section on the line B B, in which *a a a a, b b b b* are the posts standing on the parts *b g* in fig. 1. Fig. 4 is a section through C C, in which the same posts are indicated by the same letters; and fig. 5 is a plan of the domical termination of the steeple. The lines D D, E E, and F F in these figures indicate the position of the vertical section (fig. 2). The parts of the section (fig. 2) above BB will be sufficiently understood by inspection, and need not be described in detail. Fig. 6 is an elevation of the steeple. The horizontal lines A A, B B, and C C are the lines of the horizontal sections or plans, figs. 1, 3, and 4.

PLATE XLI, fig. 1.—No. 1 is a sectional elevation, and Nos. 2 and 3 are horizontal sections, of the tower of the Town-hall, Milford, Massachusetts, erected from the designs of Thomas W. Silloway, of Boston, U.S.A.

The framing is very simple. Fig. 1, No. 2, is a horizontal section or plan at the line A–B in No. 1, immediately above the floor of the first story of the tower. A A, A A, A A, A A are horizontal timbers framed between the principal posts, and strongly strapped and bolted to them, as seen in No. 1; B B, B B are dragon-pieces crossing the angles to support the posts C C, C C of the upper portion of the tower. Crossing the angles are also angle-braces, and these and the dragon-pieces are securely bolted to the horizontal timbers. No. 3 shows a horizontal section at C–D, wherein the arrangements are of much the same nature as in No. 2. The principal posts carry strong capping-pieces F F, F F, which again carry the dragon-pieces L L, L L, and are also firmly united by the angle-braces O O, O O. The use of the various struts and braces is so obvious as not to require further description.

Fig. 2, No. 1.—In this figure the principal posts are capped over at the level where the tower diminishes in diameter; and the posts for the octagonal upper portion of the tower are, as in the preceding example, carried by dragon-pieces resting on and bolted to horizontal timbers framed into the principal posts, as seen at No. 2. The arrangement of the horizontal timbers or capping-pieces for the support of the octagonal spire is shown in No. 3.

PLATE XLII.—Spire of La Sainte Chapelle, Paris.

The ancient edifice which this spire surmounts was built in 1248 to contain the relics which St. Louis brought from Palestine. It was erected from the designs of Pierre de Montreuil, the celebrated architect, who built also the castle of Vincennes. The original spire was demolished a short time before the Revolution. It was said to be a marvel of boldness and lightness. The new spire, and the scaffold used in its erection, are figured in the plate.

Fig. 1 is a section of a part of the roof and an elevation of the spire. Fig. 2 is a horizontal section or plan immediately above the roof, and shows also the disposition of the principals which support the spire.

Fig. 3 is the elevation, and figs. 4 and 5 plans of the scaffolding used in the erection of the spire.

Fig. 6.—The upper half of this figure is the half-plan at A B, fig. 1. The lower half is the half-plan on the line C D.

Fig. 7.—The upper half of this figure is the half-plan of the base of the spire immediately above the roof; and the lower half is the half-plan on the line E F, fig. 1.

Plate XLIII shows the tower and spire of the church at Grinsted-juxta-Ongar, Essex, as measured by the writer in 1899. The corresponding timbers are not of uniform cross-section. The tower is about 12 feet square, and has a post about 9 inches square at each angle, and these posts are tied together by horizontal and diagonal members, between which upright studs (about 4 × 3 inches) are framed; it is covered externally with weather-boarding of ordinary type. The beams or pole-plates connecting the heads of the four story-posts support two horizontal timbers or tie-beams (about 8 × $9\frac{1}{2}$ inches), which are halved together at their intersection on the central axis of the tower. From the point of

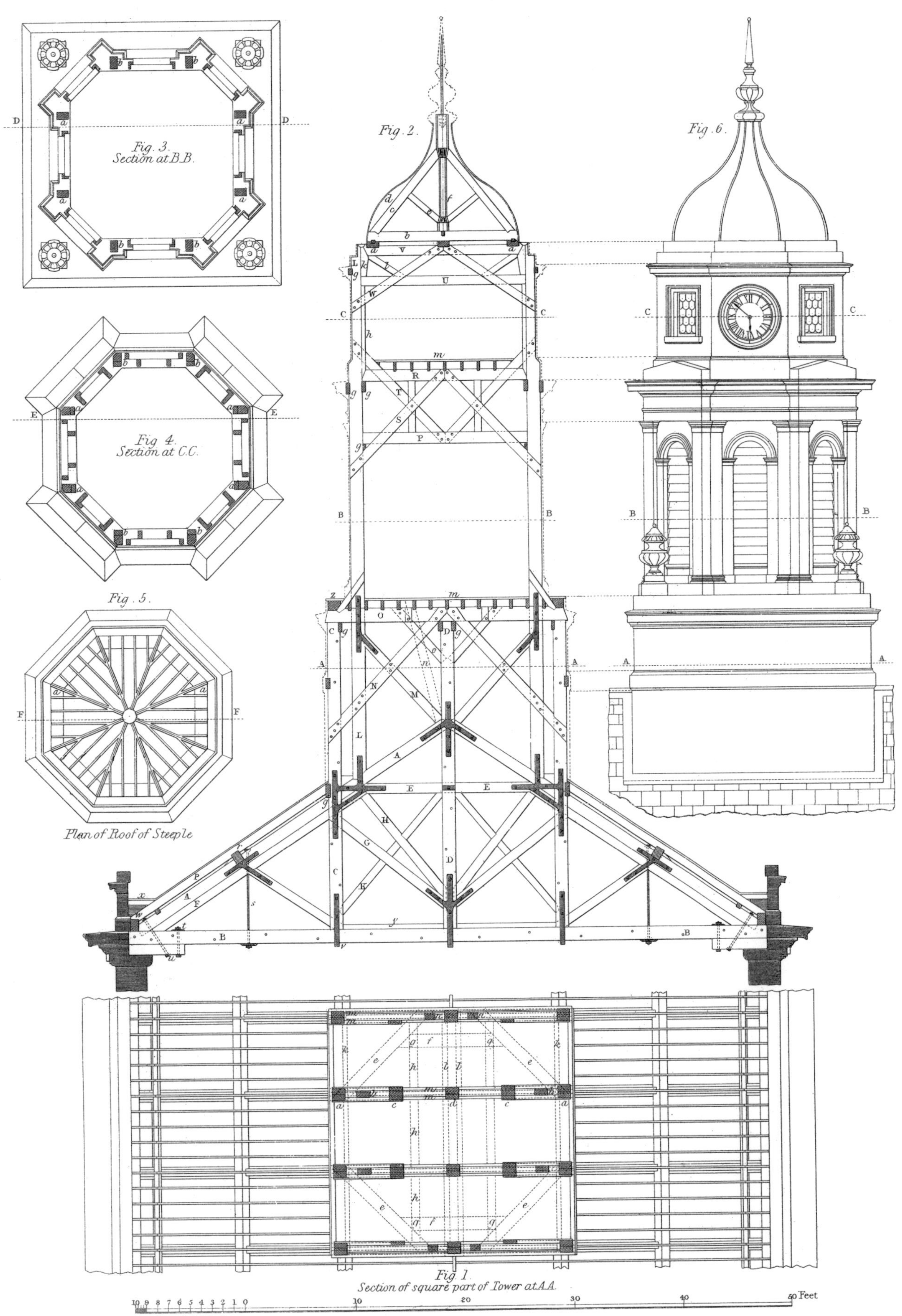

ROOF WITH TIMBER STEEPLE

intersection the central post of the spire (about $9\frac{1}{2}$ inches square at the foot and tapering slightly towards the top) is carried. Four struts ($4\frac{1}{2} \times 3\frac{1}{2}$ inches) are framed into the tie-beams and king-post. The eight hip-rafters (about 4×3 inches) are nailed to the pole-plates at the foot and secured to the king-post at the head; between them $3 \times 3\frac{1}{2}$-inch purlins are framed, and serve to support the common rafters on the four splayed sides of the octagonal spire. The common rafters are about 3×4 inches, and are strutted from the king-post by two heights of $2\frac{1}{2} \times 4$-inch horizontal struts, notched on to the rafters and nailed. The lower group of struts occupies the space shown by the brackets at C C, and the upper range that at D D. The spire is covered with oak shingles nailed to laths about 3 inches wide and $2\frac{1}{2}$ inches apart.

8. HIP AND VALLEY RAFTERS, PURLINS, DORMERS, GUTTERS, &c.

Hip-rafters.—The methods of setting out hip-rafters have been considered in Section 5 of the preceding chapter. A hip-rafter is a raking beam, supported at its foot on the wall and at its head by the ridge-piece. The upper ends of the short common rafters, known as "jack" rafters, abut against the hip-rafter and are spiked to it; if these are the only timbers supported by the hip-rafter, the load may be regarded as equally distributed. In many cases, however, purlins are framed into the hip-rafter, and the load will be partly distributed (by the jack-rafters), and partly concentrated at one or more points (by the purlins). The scantling of a hip-rafter must be calculated as that of an ordinary beam. If, however, the calculations are based on the actual length of the rafter, the dimensions obtained will be excessive. On the other hand, if the horizontal span (that is, the length of the horizontal projection of the rafter) is taken as the basis of calculation, too small a scantling will be obtained. It must not be forgotten that, while the dead load of a roof acts vertically downwards, the wind-pressure is supposed to act at right angles to the rafters. The length to be adopted in calculating the strength must therefore lie between the actual length of the rafter and the length of its horizontal projection.

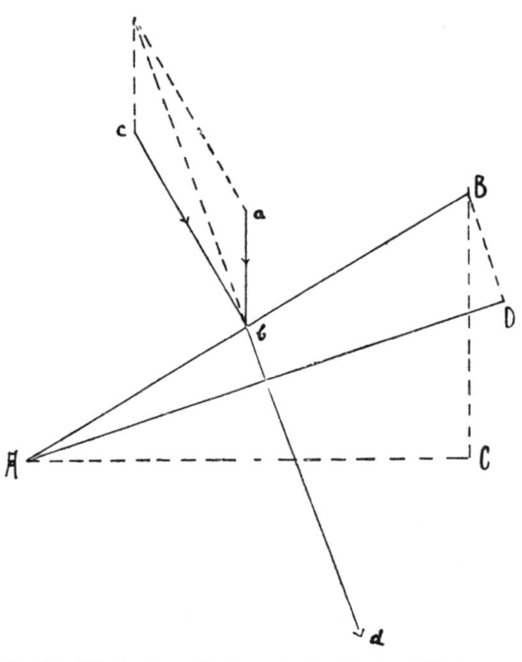

Fig. 786.—Effective Span of Rafter under Dead Load and Wind-pressure

Thus, in fig. 786, A B is the actual length of a rafter, and A C its horizontal projection or span. Let *a b* be the graphic representation of the dead load, and *c b* that of the wind-pressure; then *bd* is the resultant. From A draw A D at right angles to *bd*, and from B draw B D parallel to *b d*. A D is the span to be used in calculating the strength of the rafter. If the hip-rafter is supported from the attic floor by means of ashlaring, it becomes a continuous beam, and the scantling can be reduced. It is obvious that a hip-rafter exerts an outward thrust on the wall. The method of securing the foot of the rafter in order to distribute the thrust over as large an area of wall as possible is therefore important. The best arrangement is to frame it into a "drag-baulk" or "dragon beam", this in turn being secured to an angle-tie, as shown in fig. 787. In large roofs the hip-rafters may be supported by one or more trusses carried across the room, as in the City Hall at Glasgow (p. 42), or a half-truss may be formed under each hip, resting at one end on the angle of the wall and at the other end on a strong truss carried across the room, as in the Waterloo Rooms, Glasgow (p. 41).

Valley-rafters are stressed in a similar manner to hip-rafters, and the scantling must be calculated in the same way. The jointing at the foot is not, however, so important, as the valley-rafter abuts against a re-entrant angle in the walling, and the two walls meeting thus usually furnish ample resistance to the thrust. Where one roof intersects a loftier roof, the heads of the two valley-rafters must be carried by a purlin, or (where there are no purlins) one of the valley-rafters must be continued to the ridge, and the other framed into it, as shown in the lower part of fig. 788. Similarly, in hipped roofs it is sometimes necessary to continue a hip-rafter to the wall (as in the upper part of the fig.).

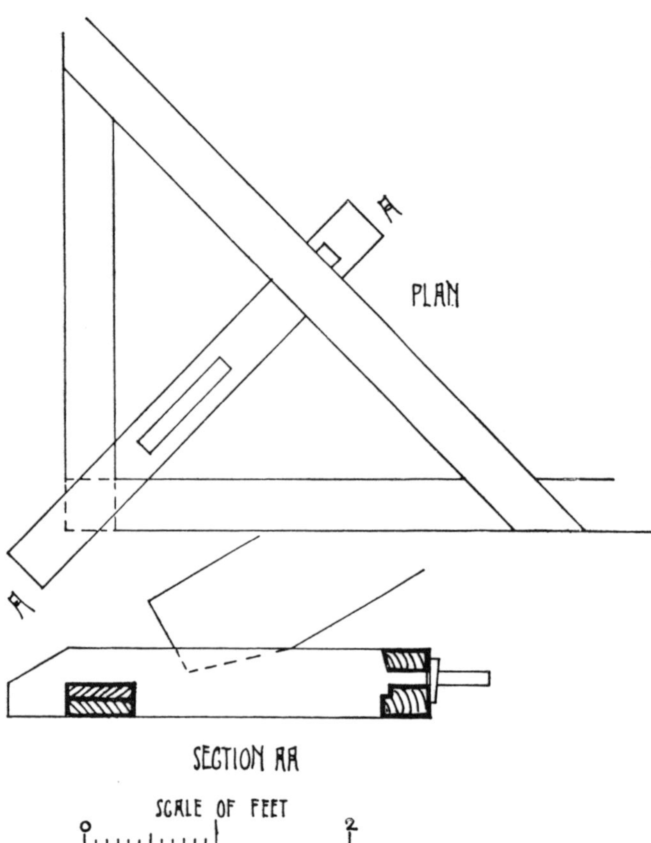

Fig. 787.—Joint at Foot of Hip-rafter

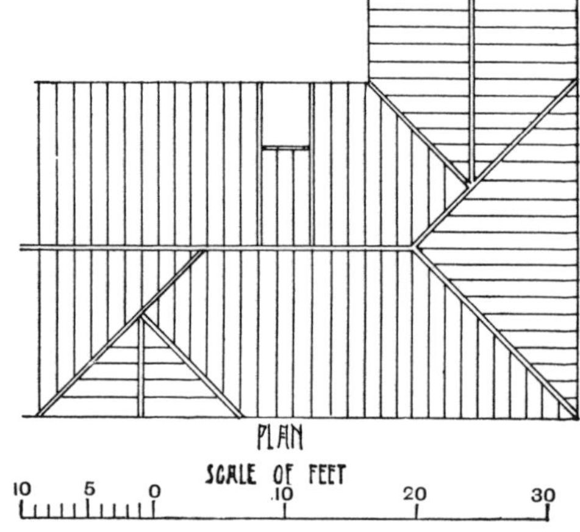

Fig. 788.—Plan of Roof with Valley-rafters and Hip-rafters

Purlins are usually carried by walls or roof-trusses. They may also be framed into hip and valley rafters with a single or double tenon, and secured with spikes. Opinions differ as to the best method of placing purlins with reference to the slope of the roof. As a rule, two sides are parallel to the slope and two at right angles to it, but if the depth greatly exceeds the breadth, this arrangement is not suitable for roofs of quick pitch. In many districts the purlins are placed with two of the sides vertical, and a triangular piece is sawn off the top to the required roof-slope; such purlins are sometimes known as "wavers". Probably the most correct angle will be found in the manner shown in fig. 786; in this example two of the sides will be parallel to the resultant *b d*.

Dormers.—In roofs with purlins there is not much difficulty in constructing dormers, as the plates and valley-rafters of the dormers can be supported on the purlins of the main roof. Where only common rafters are used, they must be trimmed after the manner shown in fig. 788. In America, it is a common practice to nail two or three rafters together to give the necessary strength to the trimming-rafters at the sides of the opening. The trimmer or header at the top of the opening is framed into the trimming-rafters. Fig. 789 shows the framing of a simple dormer in a room where the rafters are supported by ashlaring. The sides of the dormer are formed by vertical studs, those marked *a* extending from the floor to the plate *b*; these studs should be notched out about 1 inch over the trimming-rafter and securely nailed, as this stiffens both the dormer and the trimming-rafters. Above the main roof the sides or "cheeks" of the dormer may be covered with boarding and lead, or tile-hung, or finished in other ways. Sometimes the cheeks are fitted with window-frames.

Six designs for dormers are given in fig. 790. No. 1 is hipped, and has a pitch suitable

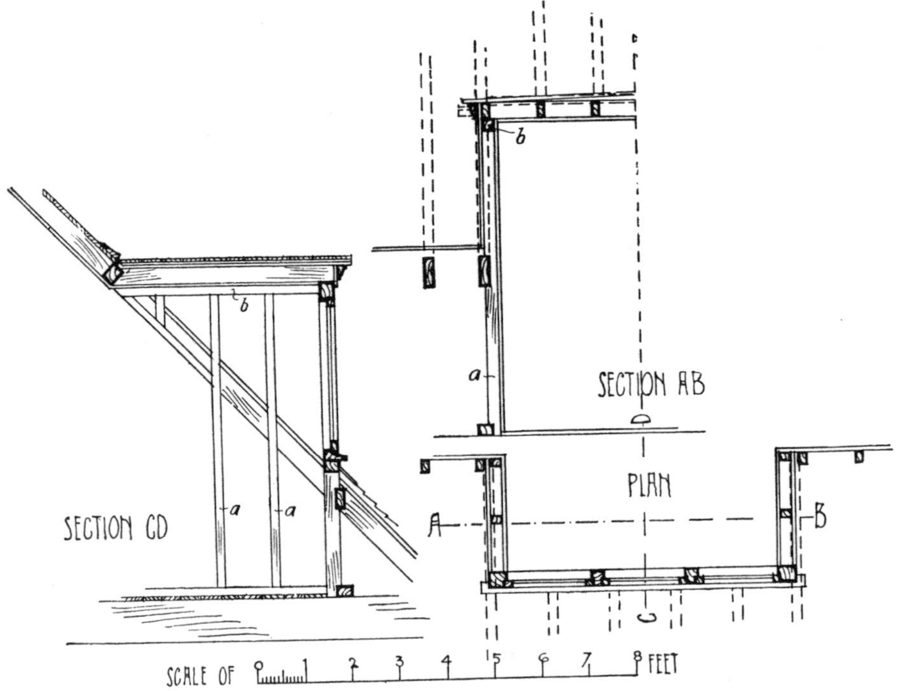

Fig. 789.—Dormer in Common-rafter Roof, with Ashlaring

for slating; Nos. 2 and 3 are for the lower slopes of curb roofs, and are intended to be covered on the top and cheeks with boarding and lead; Nos. 4, 5, and 6 are also designed for lead. Other designs will be given in the section on joinery.

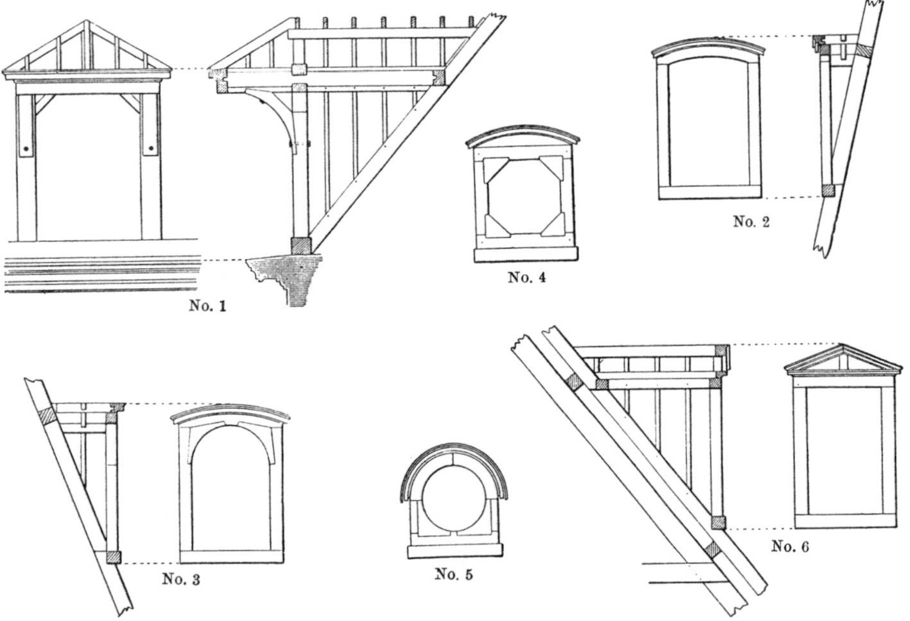

Fig. 790.—Six Designs for Small Dormers

Gutters.—Wooden gutters covered with lead are often used in the hollows of **M**-shaped roofs, and behind parapets and over valley-rafters. Fig. 791 gives the plan of a tapering gutter 30 feet long in the hollow between two span-roofs, and a transverse section to a larger scale, together with a detail of one of the drips. A1, A2, A3 are the three "flats", sloping not less than $1\frac{1}{2}$ inch in 10 feet; B1 and B2 are the two drips, 3 inches deep (in no case should a drip be less than 2 inches deep); C is a wood roll dividing the uppermost flat longitudinally, as the lead would otherwise be of great width, and therefore likely

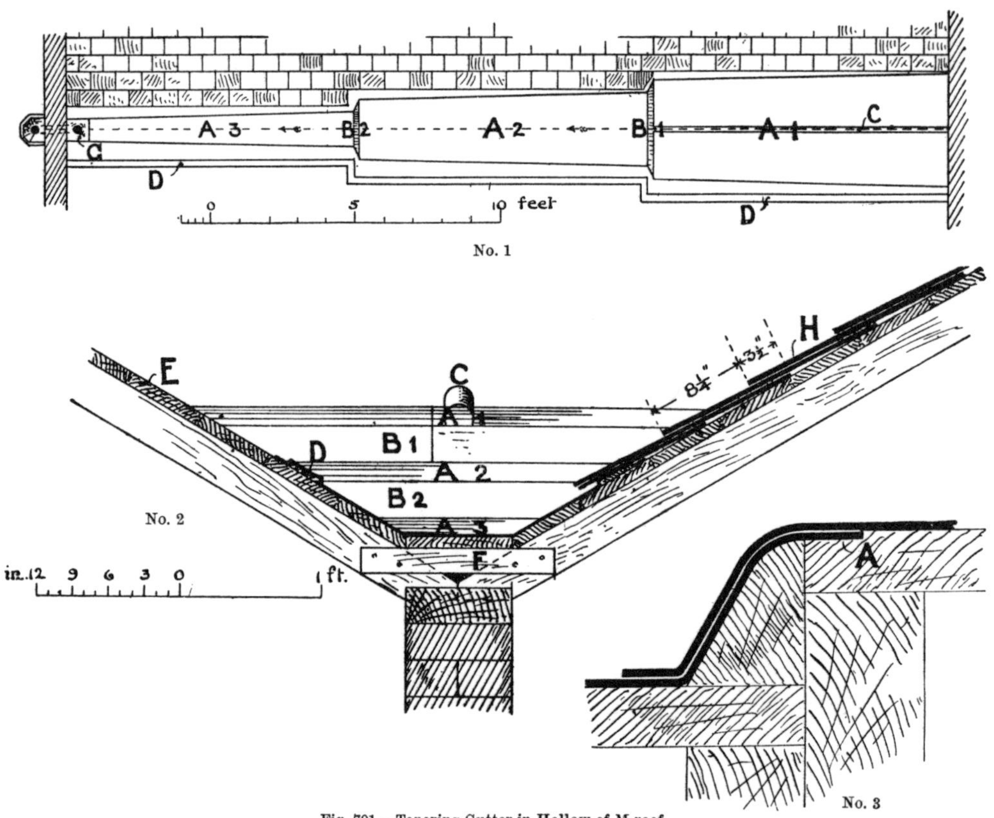

Fig. 791.—Tapering Gutter in Hollow of M-roof

to crack; D is the tilting fillet for the slates, thickest at the lower edge; E, the roof-boarding covered with felt or waterproof paper; F, the gutter-bearer; G, the cesspool or small sunk box (from 8 to 12 inches square and 6 to 8 inches deep) formed at the outlet end of the gutter, and fitted with a lead outlet-pipe leading through the wall to the head of the rainwater-pipe (the boarding around the circular opening for the outlet-pipe should be chamfered to allow the plumber to make a water-tight joint without forming a ridge around the opening); at H the slates are shown, 20 inches long, laid to a lap of $3\frac{1}{2}$ inches, and nailed directly to the roof-boarding. No. 3 is an enlarged section of one of the drips, the boards being rebated at A to the thickness of the underflashing, and a rounded fillet being inserted to prevent the cracking of the lead.

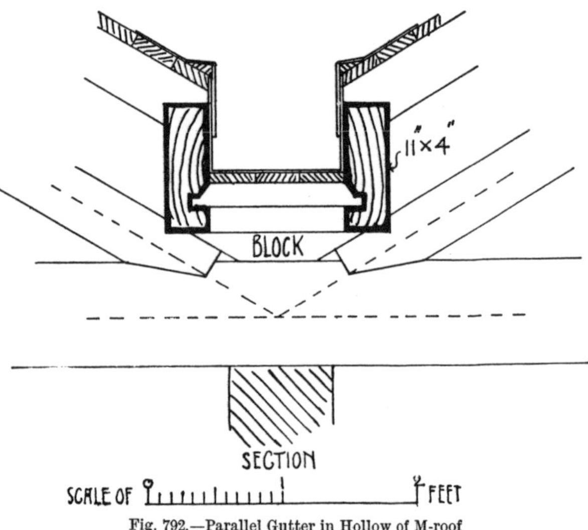

Fig. 792.—Parallel Gutter in Hollow of M-roof

Parallel gutters (also known as "box" or "trough" gutters) may be used in a similar position to the last, if pole-plates can be introduced to carry the feet of the common rafters. Fig. 792 shows such a gutter in section. The bottom of the trough must be laid with falls and drips as in the tapering gutter. The width of the trough ought not to be less than 12 inches, nor the depth at any point less than 6 inches. The two 11 × 4-inch pole-plates are carried by the principal-rafters and the block or packing between them; gutter-bearers are framed into the plates, to receive the boarding; and tilting-fillets are nailed along the edges of the gutter.

A tapering parapet-gutter is shown in cross-section in fig. 793. The construction is similar to that of fig. 791. Snow-boards nailed to arched bearers are shown at A; the boards are from 2 to 3 inches wide, and about $\frac{1}{2}$ inch apart, so as to allow the passage of water but not of snow. The overflow-pipe B serves as an escape for water when the outlet-pipe from the cesspool is choked. Gutters of similar construction are formed along the backs of chimney-stacks. Parallel parapet-gutters may be formed by means of a pole-plate, after the manner shown in fig. 792.

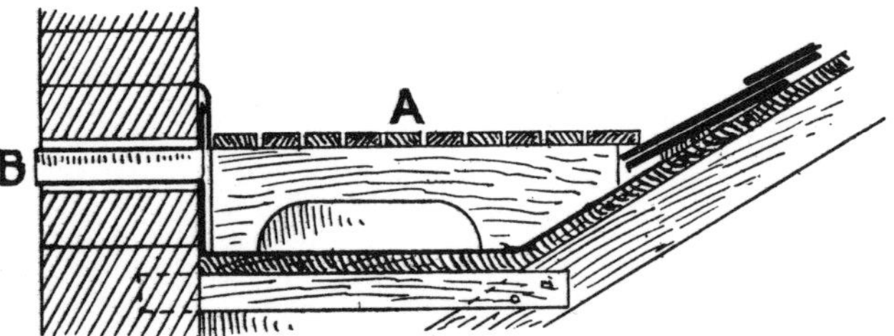

Fig. 793.—Tapering Parapet-gutter

Valley-gutters to receive lead flashing are made by laying a board about 9 inches wide and $\frac{3}{4}$ inch thick on each side of the angle, and nailing it to the common rafters. A tilting fillet should be nailed along each board near the outer edge to give the necessary tilt to the slates, and to prevent water passing over the edge of the lead.

Lead-flats.—A lead-flat for a bay-window is shown in fig. 794, No. 1 being the half-plan, and No. 2 the section on the line B B to a larger scale. In the plan F F are the tapering joists, G G the upper joists or purlins, C C the rolls, D the gutter, and E the drip in the gutter. In this case there is no cesspool, the bottom of the gutter at the outlet end being slightly dished, and the outlet-pipe being taken from the end of the gutter instead of from the bottom. The rolls are usually 2 or $2\frac{1}{2}$ inches in diameter, the cross-section being about three-quarters of a circle; sometimes the sides are parallel and the top rounded. The latter shape is preferred for zinc roofs.

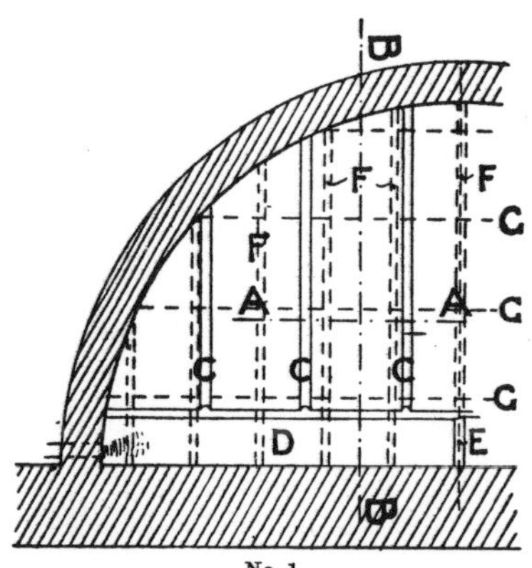

No. 1

Roll-ridges.—Wood rolls are required for ridges and hips when these are to be covered with lead. They may take the form shown in fig. 795, the rectangular packing being nailed to the ridge or hip-rafter, and the roll being nailed to the packing.

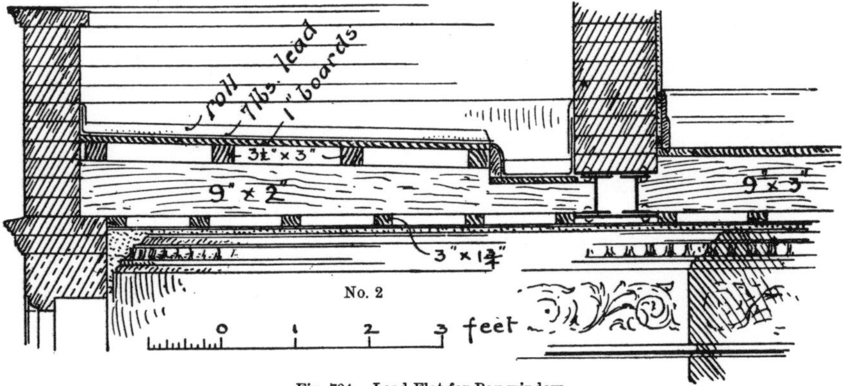

Fig. 794.—Lead Flat for Bay-window

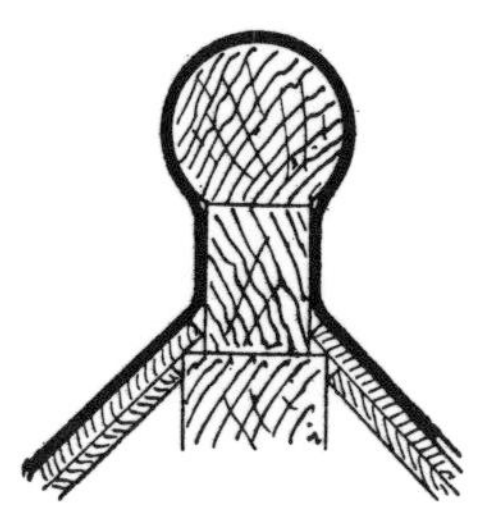
Fig. 795.—Wood Roll for Ridge

Or double-pointed nails may be used, the lower point being driven through the packing into the ridge-piece, and the roll being then laid on the upper point and driven home.

Eaves.—The angle in a curb roof must also be prepared for lead-flashing. One method is shown in fig. 796. Sometimes an iron or wood eaves-gutter is fixed to the feet of the upper rafters or to a face-board, and the water from the upper slope is collected in this gutter and conveyed by rainwater pipes or channels to the lower gutter. This arrangement reduces the risk of rain being driven through the eaves.

Wood eaves-gutters are not used as frequently as in former years, but in many situations they are more durable than iron if the wood is well-seasoned and free from sap and shakes. Thus, iron is soon corroded by the fumes from the sulphur-stoves of blanket manufactories, and by the ammonia from the urine of sheep, pigs, and other animals. In its simplest form the wood eaves-gutter is rectangular in cross-section, with a semicircular hollow formed on the upper side. To improve the effect the trough itself may be moulded on the lower front edge, and an additional mould planted on the upper edge, as in fig. 797. For more ornamental work, brackets and other members may be introduced, as shown in fig. 730, No. 2. If joints are required in the length of the gutter or at mitres, they should be formed by slightly rebating the hollow in each portion for a distance of about 2 inches from the joint, and inserting a piece of lead or copper, bedded in red and white lead and secured with small clout-nails. Nozzles or "drops" should be formed with lead or copper funnels, flanged at the top, and let into rebates in a similar manner. As a rule, gutters of this kind are tarred on the upper surfaces and painted on the other sides, or the upper surface may be entirely covered with lead.

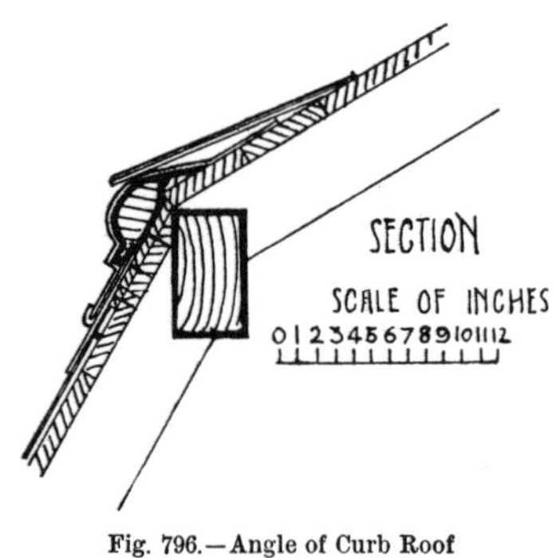

Fig. 796.—Angle of Curb Roof

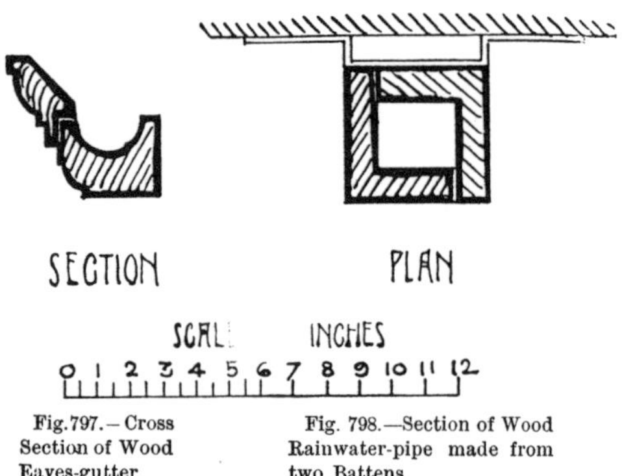

Fig. 797.—Cross Section of Wood Eaves-gutter

Fig. 798.—Section of Wood Rainwater-pipe made from two Battens

Wood rainwater-pipes are less frequently used than wood eaves-troughs, but are still preferred by many persons. They may be constructed of four pieces ($\frac{7}{8}$ or $1\frac{1}{8}$ inch thick) nailed or screwed together, the abutting surfaces being first covered with red and white lead, and the four surfaces which will form the interior of the pipe being painted or tarred. Sometimes the front piece is made slightly convex or moulded. Another method consists in cutting a rectangular piece about $3\frac{1}{4} \times 1\frac{3}{4}$ inches out of each of two $4\frac{1}{2} \times 3$ battens, and then nailing the two battens together, as shown in fig. 798. This is somewhat wasteful of timber, but reduces the number of joints from four to two.

Wood rainwater-pipes are usually fixed close to the wall, and the backs ought to be painted two or three coats before this is done. If wrought or cast iron ears of the shape shown in fig. 798 are used, the back can be painted periodically and the wood preserved for a longer period.

CHAPTER VII

FLOORS AND CEILINGS

Floors are the horizontal partitions which divide a building into stages or stories. The timbers which enter into their composition are bridging-joists, trimmers and trimming-joists, binding-joists, girders, ceiling-joists, and the boards which form the platform. All these, except the last-mentioned, are comprehended under the term "naked flooring".

When the bearing between the points of support is not great, bridging-joists alone are used to support the flooring-boards. They are laid across the opening or void, and rest on the wall at each end. A piece of timber, called a wall-plate, is often interposed

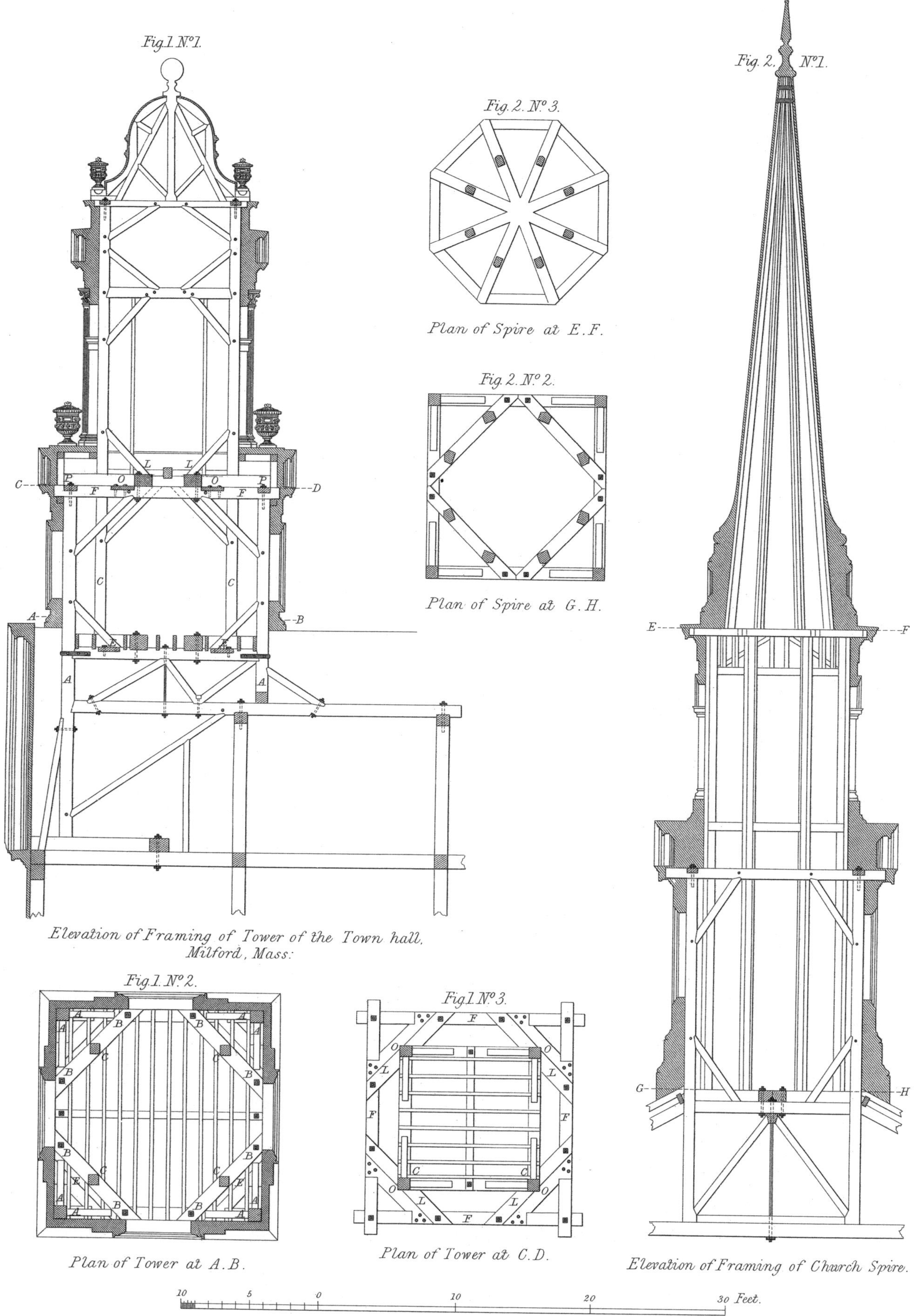

TIMBER STEEPLES—AMERICAN EXAMPLES

between the ends of the joists and the wall, to equalize their bearing. If the wall is of the same thickness above and below the floor, the wall-plates must be built into the wall, and in such a position are apt to decay, and to be compressed to some extent by the weight of the wall above. They are also a source of danger if a fire occurs. For these reasons the use of such plates or bond-timbers is prohibited by many public authorities.

A floor of bridging-joists is called a single-joisted floor. When the bearing is long, the joists, from their elasticity, bend under a moving weight, and thus disturb the ceiling below. The deflection is considerably diminished by placing underneath them stronger timbers, called binders or binding-joists. This construction is called a double floor. When it is calculated that the bearing will exceed the limit of strength of the binding-joists a third mode of construction is adopted, in which large timbers, called girders, are introduced to support the binding-joists. This construction is called a framed floor. Steel girders are now generally used instead of wood for large spans and heavy loads, and framed timber floors are therefore seldom constructed.

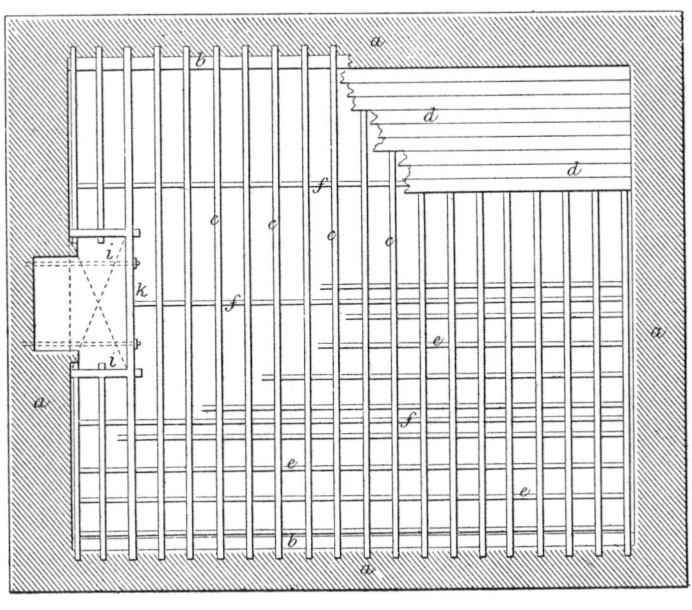

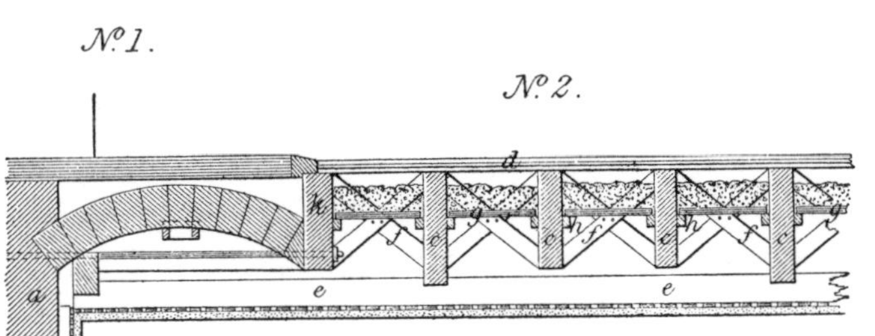

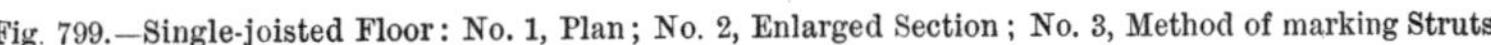

Fig. 799.—Single-joisted Floor: No. 1, Plan; No. 2, Enlarged Section; No. 3, Method of marking Struts

Bridging-joist or Single-joisted Floors (fig. 799).—No. 1 is the plan of an apartment: *a a a a* are the walls, *b b* the wall-plates, *c c c c*, &c., the bridging-joists, *d d* part of the flooring-boards. The bridging-joists are usually placed from 10 to 12 inches apart; their scantling is dependent on their length, their distance apart, and the weight they have to carry. Sometimes every third or fourth bridging-joist is made deeper than its fellows, and the ceiling-joists are then fixed to them only. This has the advantages of preventing sound passing so readily, and making the ceiling stand better.

When some joists would, from their position, run into a fireplace or flues in a wall, it is improper to give them a bearing there. In the case of the floor, fig. 799, two short timbers *i i*, called "trimmers", are introduced—one on each side of the hearth, with one end resting in the wall, and the other framed into the third joist, which is known as a trimming-joist. The inner ends of the two first bridging-joists are framed into the trimmers. We may define a trimming-joist as a floor-joist parallel to the bridging-joists, and sustaining one or more trimmers, which form an angle with it—generally a right angle. Trimmers and trimming-joists are made thicker than the others, according to the number of joists dependent on them for support. The hearth rests on a brick arch turned between trimming-joist and wall. The curve of the arch may be as shown, or a half-arch may be used, the highest part being against the trimming-joist. Instead of a brick arch, curved steel plates are now sometimes used, with a vertical flange nailed or screwed to the trimming-joist, the plates being afterwards covered with concrete to receive the hearth tiles; or concrete alone is used, resting on fillets nailed to the sides of the trimmers and trimming-joists, and supported while setting by temporary sheeting; wood grounds to receive the plasterers' laths are fixed before the concrete is deposited.

No. 2 shows a section through the hearth and floor at right angles to the direction of the joists: *c c* are the bridging-joists, *d* the edge of one of the flooring-boards, *e e* the side of a ceiling-joist. The ceiling-joists cross the bridging-joists at right angles, as seen at *e e e*, No. 1, and are notched up to every third joist (which is deeper in this case than the other joists) and fastened with nails. For cheap work the ceiling-joists are omitted, the plasterers' laths being nailed directly to the floor-joists, which are all of the same depth.

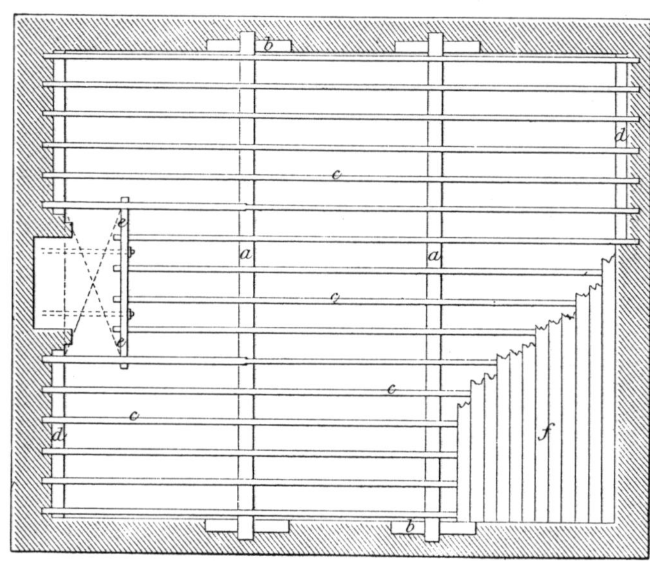

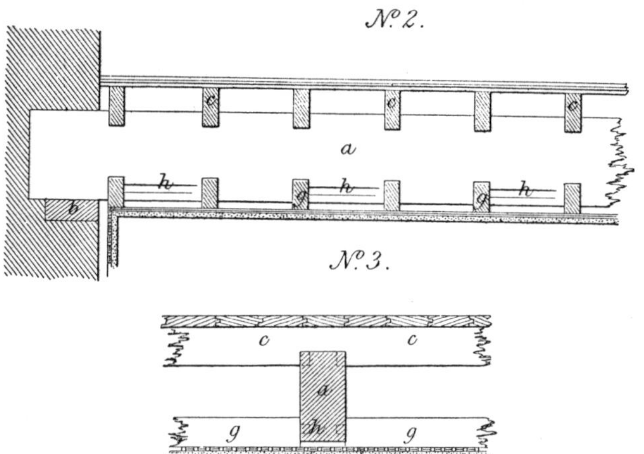

Fig. 800.—Double Floor (for scales see fig. 801)

When the bearing of single joists exceeds 8 feet they should be strutted between, to prevent their twisting, and to give them stiffness. When the bearing exceeds 12 feet, two rows of struts are necessary; and so on, adding a row of struts for every increase of 4 feet in the bearing. Strutting does not increase the *total* strength of the floor, but is of service in distributing a concentrated load on one joist over the adjacent joists. It thus reduces the risk of any one joist being unduly deflected and of the plastered ceiling beneath being cracked.

There are three modes of strutting employed. The first and most simple is to insert a piece of board, nearly of the depth of the joists, between every two joists, so as to form a continuous line across. The struts should fit rather tightly, and be nailed to keep them in position. The second mode is to mortise a line of stout pieces into the joists in a continuous line across, but the mortises materially weaken the joists. The third mode is represented in the section, No. 2: *f f* are double struts, of pieces from 3 to 4 inches wide and $1\frac{1}{2}$ inch thick, crossing each other, and nailed at the crossing to each other, and at their ends to the joists. The struts should be cut at their ends to the bevel proper for their inclination. To save the trouble of boring holes for the nails, two slight cuts are made at each end with a wide-set saw, and the strut is nailed through these with clasp-nails. Of the three modes the last is the best. The method of finding the exact angle of the end of the strut is to draw two parallel lines across and at right angles to the joists along the whole length of the floor, as at *a a*, *b b* (No. 3), the lines being at a distance from each other equal to the depth of the joists; then lay the scantling so that its opposite sides intersect the angles formed by the lines and the joists; the line described by each joist on the scantling above it will be the required angle of the end of the dwang. In No. 1, *fff* show three lines of struts.

In order to effectually prevent the passage of sound from one story to another, a second

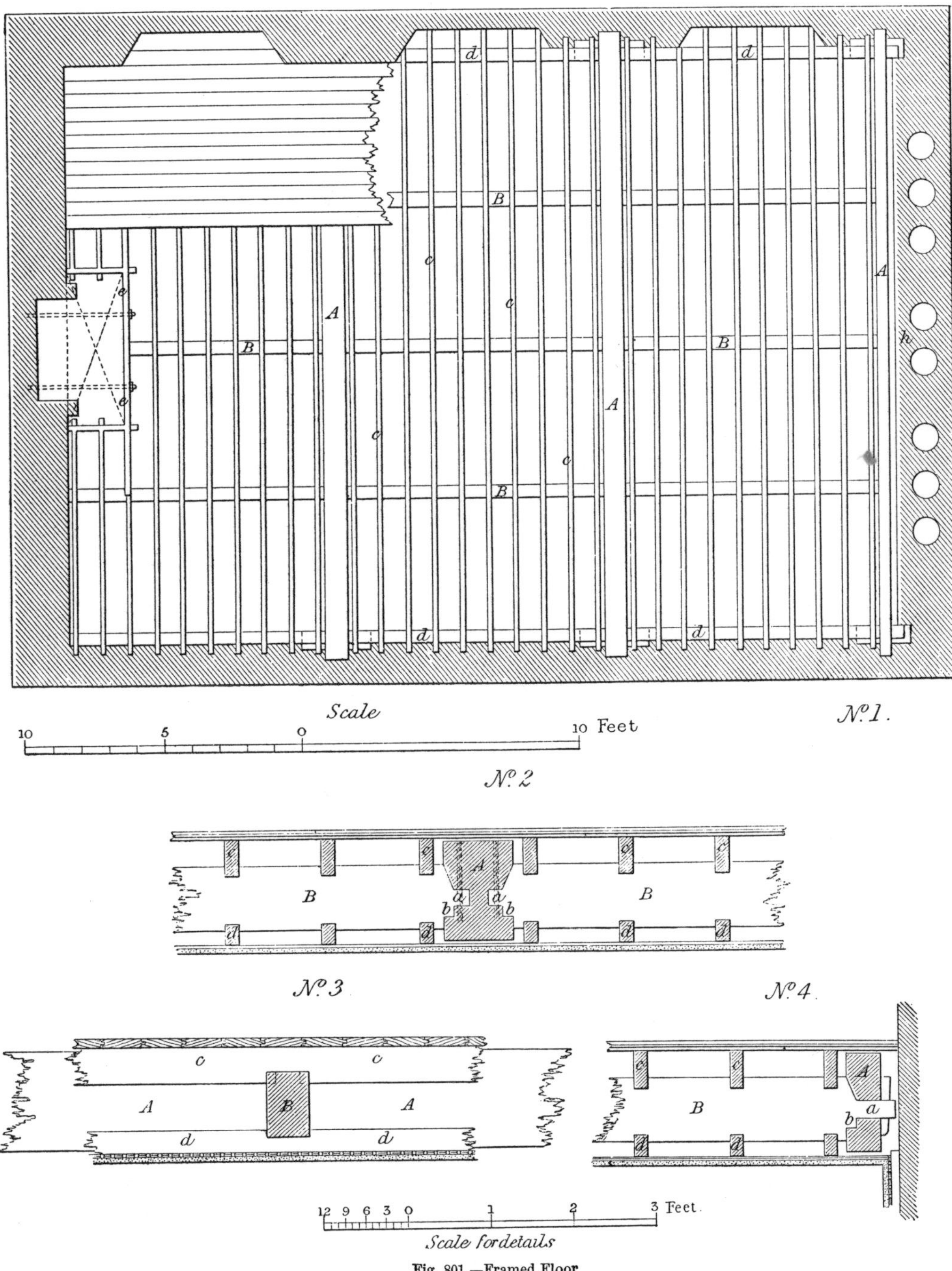

Fig. 801.—Framed Floor

floor of rough boarding is sometimes inserted between the joists, and covered with some non-conductor of sound—the usual composition, which is called in England *pugging,* and in Scotland *deafening,* being a mixture of lime mortar and clean light ashes. A much better material, however, is silicate cotton or slag-wool. This sound-boarding, or deafening-boarding, as the secondary floor is called, is supported on fillets nailed to the sides of the joists. It is shown in the section, fig. 799, No. 2, where *g g* is the boarding and *h h* the fillets. Along the joist next the wall a fillet is nailed, so as to fill up the space between the joist and the wall, and admit of the pugging being used there to effectually stop communication.

Double Floor, or Floor with Binding-joists (fig. 800).—No. 1 is a plan of this kind of floor: *a a* are the binding-joists, having their ends resting in the walls, but with templates *b b*, which are short pieces of timber or stone, interposed to lengthen their bearing; *c c* are the bridging-joists, *d d* the wall-plates, *e e* is a trimmer opposite the fireplace, framed at the ends into the two trimming-joists, and *f* part of the boarding.

The section, No. 2, shows the connection of the parts: *a* one of the binding-joists, *c c* the bridging-joists, cogged over the binding-joist, and *g g* the ceiling-joists. Where the

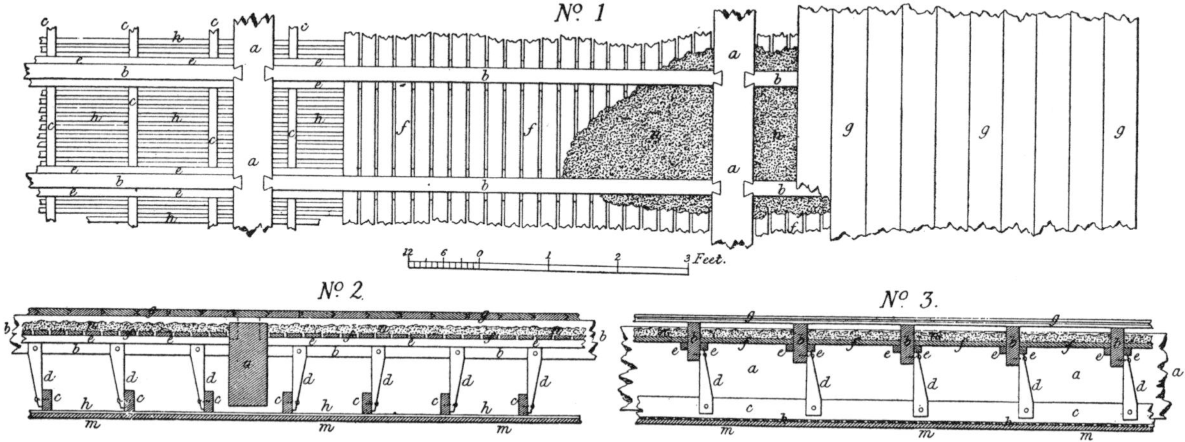

Fig. 802.—Floor with suspended Ceiling

saving of depth in the framing is an object, the ceiling-joists are framed into the binding-joists by a chase-mortise, as at *h* in the same figure, or forked on to fillets nailed to the sides of the binding-joists (see fig. 621). No. 3 is a section of the same floor parallel to the direction of the bridging and ceiling joists. The same letters of reference apply to the same parts in both figures.

Framed Floor (fig. 801).—No. 1 represents the plan: A A are the girders, with their ends bearing on templates in the walls; B B B are the binding-joists, and *c c* the bridging-joists; *d d* the wall-plates; *e e* the trimming-joist at the fireplace. As the wall *h* contains flues along nearly all its length, the binding-joists do not rest in it, but are framed into an additional girder A. In this case the tenon passes through the mortise, and is keyed on the other side, as shown in section in No. 4, in which A is the wall-girder and B the binding-joist.

No. 2 is a section through the girder, showing the manner of framing the binding-joists into it: A the girder, B B the binding-joists, *c c* the bridging-joists, *d d* the ceiling-joists, which are notched and nailed to the binding-joists. No. 3 is a section through the floor at right angles to the last section; in it the same letters refer to the same parts. In framing the binding-joists into the girders a "tusk-tenon" is used, as seen in the sections Nos. 2 and 4. It is necessary to take great care in fitting the bearing parts to the corresponding parts of the mortises.

It is a good practice to saw girders down the middle, and to reverse the ends and bolt the halves together with the sawn sides outwards, and with slips between to admit a circulation of air. By this means the heart of the timber can be examined, and the beam be rejected if unsound; the timber being reduced to a smaller scantling also dries more readily, and is rendered less liable to decay; and as the butt and top of a tree are rarely of the same strength, the girder must be improved by the process, which tends to equalize its strength throughout. Where additional strength and stiffness are required without increase

Fig. 1.

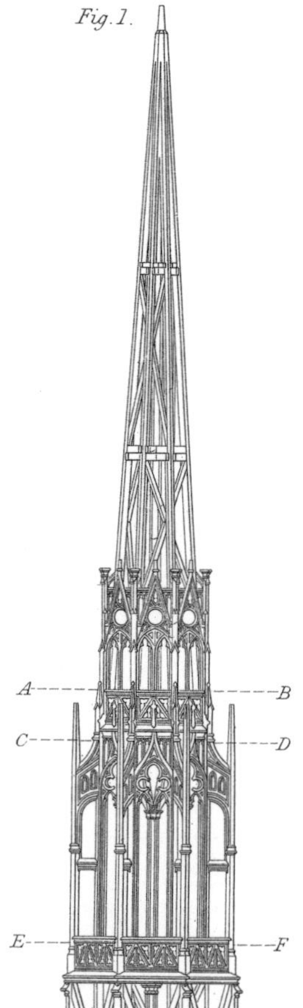

Elevation of Spire, and part of Roof.

Fig. 6.

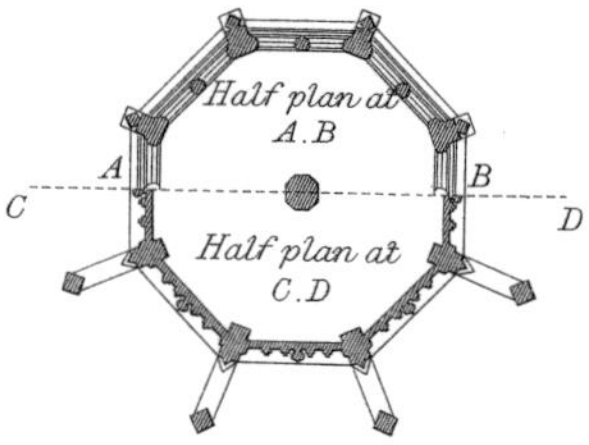

Fig. 7.

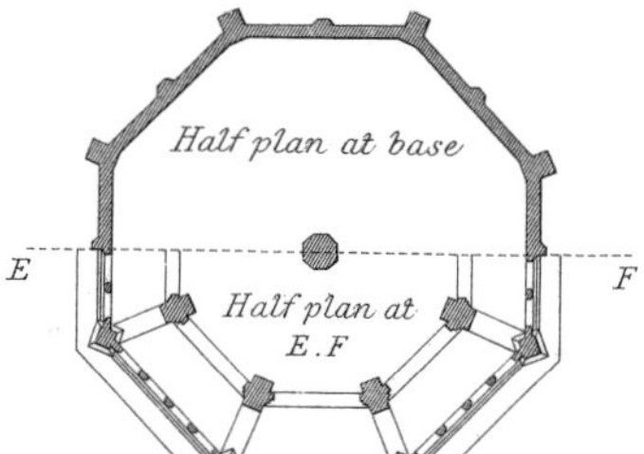

Fig. 3.

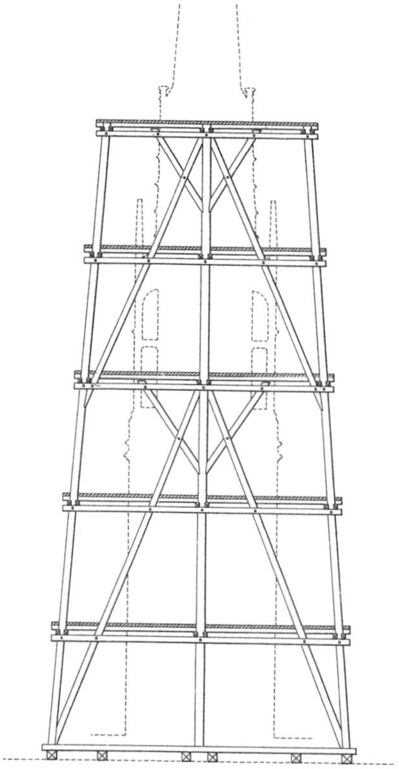

Elevation of one side of Scaffolding.

Fig. 4.

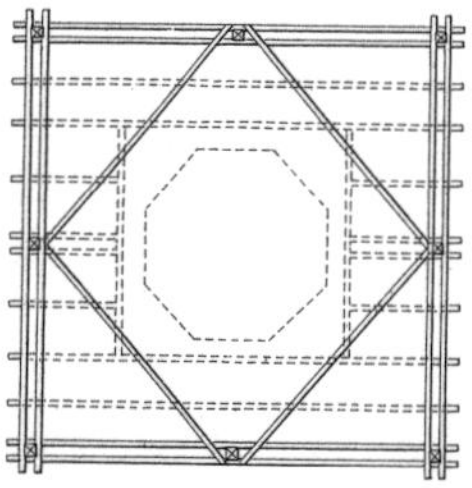

Plan of upper part of Scaffolding.

Fig. 2.

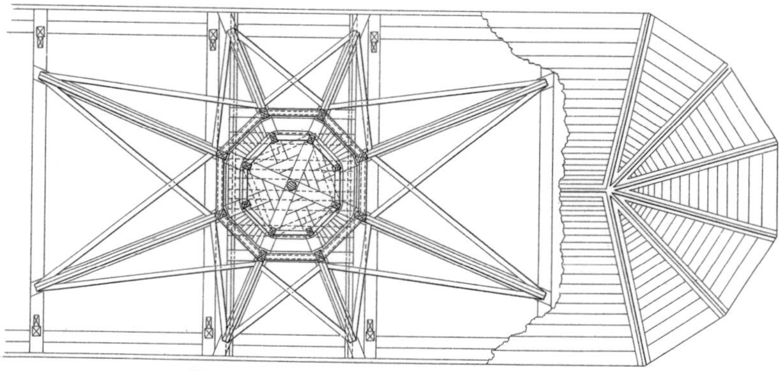

Plan of base of Spire, and part of Roof.

Fig 5.

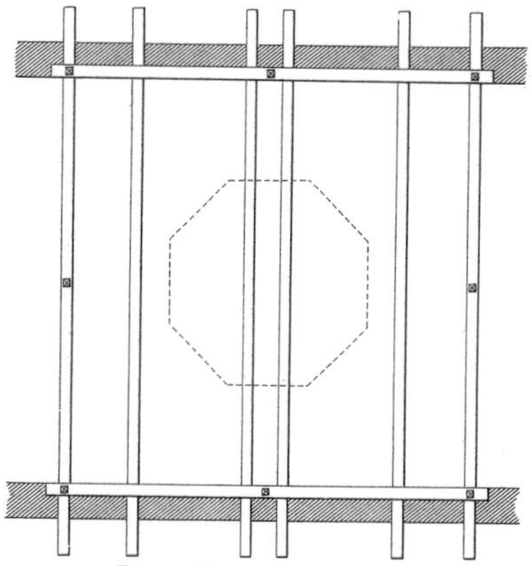

Plan of base of Scaffolding.

Scale for Figs: 1, 2, 3, 4, 5.

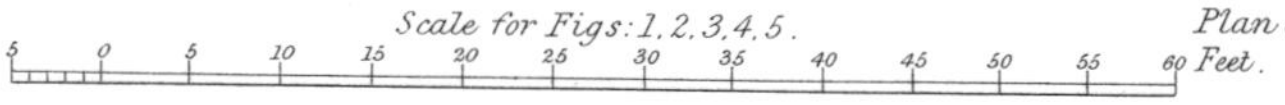

Scale for Figs: 6, 7.

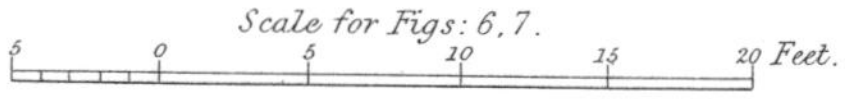

TIMBER SPIRE AND ROOFING OF LA SAINTE CHAPELLE, PARIS

WITH DETAILS OF THE SCAFFOLDING USED IN CONSTRUCTION

of depth, an iron flitch-plate may be introduced between the two timbers; this is often used in the case of long trimmers. For the strength of flitched beams see pp. 394–396, vol. i.

Variations in the Modes of Constructing Floors.—In framed floors, especially in Scotland, binders are frequently omitted, the girders are more numerous, and the bridging-joists are either notched down on them if the space will admit, or tenoned into them if otherwise. The ceiling-joists, too, in place of being notched or tenoned, are suspended to the bridging-joists by small straps of wood. Thus the separation between the floor and the ceiling is more complete, and sound is less readily transmitted. In fig. 802, No. 1 is the plan, No. 2 a transverse section across the direction of the girders, and No. 3 a longitudinal section, at right angles to the last, of such a floor. The same letters refer to the same parts in all three figures: *aa*, girder; *bb*, bridging-joists dovetailed into the girders; *cc*, ceiling-joists hung to the bridging-joists by the straps *d d*; *e e*, fillets for the support of the sound-boarding or deafening-boards, nailed to the sides of the bridging-joists; *ff*, sound-boarding lying loosely on the fillets *e e*; *g g*, flooring-boards grooved and tongued; *h h*, ceiling-laths; *m m*, plaster ceiling; *n n*, pugging or deafening. Sometimes steel angles are bolted to the sides of the beams to receive the ends of the joists, the latter being fixed to the beams by skew-nailing.

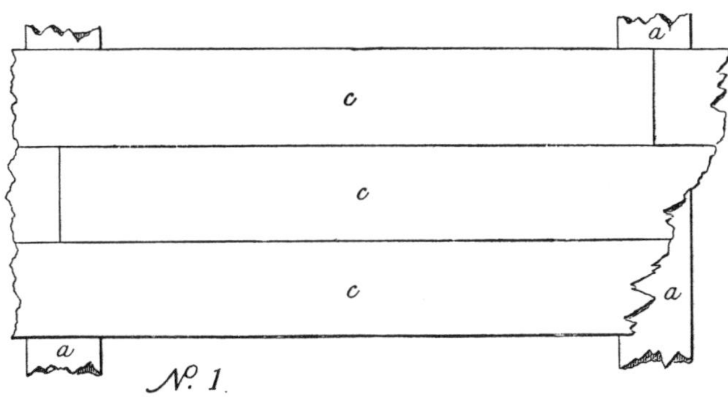

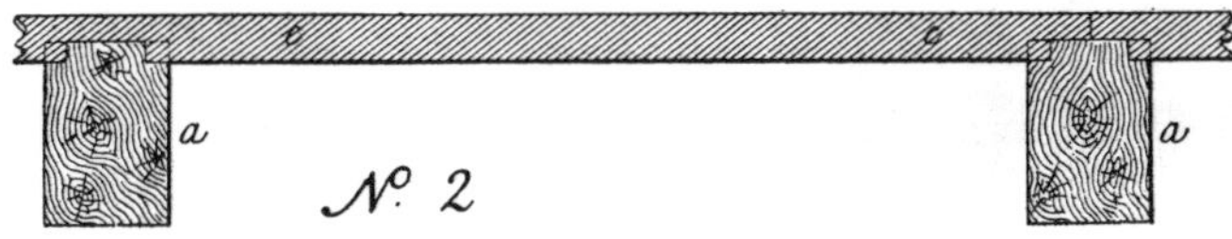

Fig. 803.—Plan and Section of Plank Floor

In fig. 803, No. 1 is the plan, and No. 2 a section of part of a warehouse floor composed of girders *a a a*, supporting the floor of planks *c c c*. The planks are shown to be cogged on to the girders, but this is unnecessary. They are generally laid across the tops of the girders, and merely secured with nails. The edges of the planks are united by slip-feathers of wood or wrought-iron.

The ends of joists on ground-floors ought to be carefully protected from damp. Two arrangements are shown in fig. 804. In No. 1 the floor is above the external ground; A is the concrete foundation, B the brick footings, C part of the footings carried up to form a set-on to receive the wall-plate, D the damp-course of perforated stoneware, E the wall-plate, F the floor-joist, G the floor-boards, and H the concrete ground-layer. Ample ventilation is necessary to prevent dry-rot, and can be secured by a ventilating damp-course, or by air-grates near the angles of the walls. In No. 2 the floor is below the external ground, and the wall is built hollow; A and B are the two ventilating damp-courses, and C the bonding-blocks tying the two parts of the wall together.

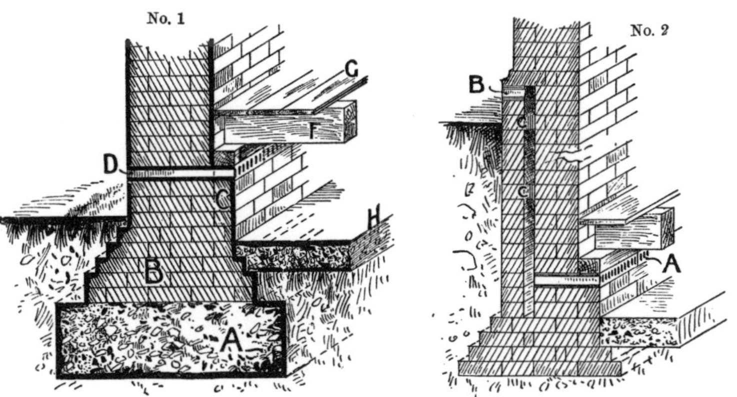

Fig. 804.—Method of supporting Ends of Ground-floor Joists

Warehouse floors are often constructed with timber beams, supporting the floor-joists and supported by cast-iron pillars (fig. 805) or by steel stancheons. The joists may have dovetailed ends fitting into corresponding notches in the beams, but these notches weaken the beams, and iron stirrups may with advantage be used instead, as shown in fig. 806. The upper pillar (fig. 805) must not rest on the beams, but on a "strider" or "beam-box", which

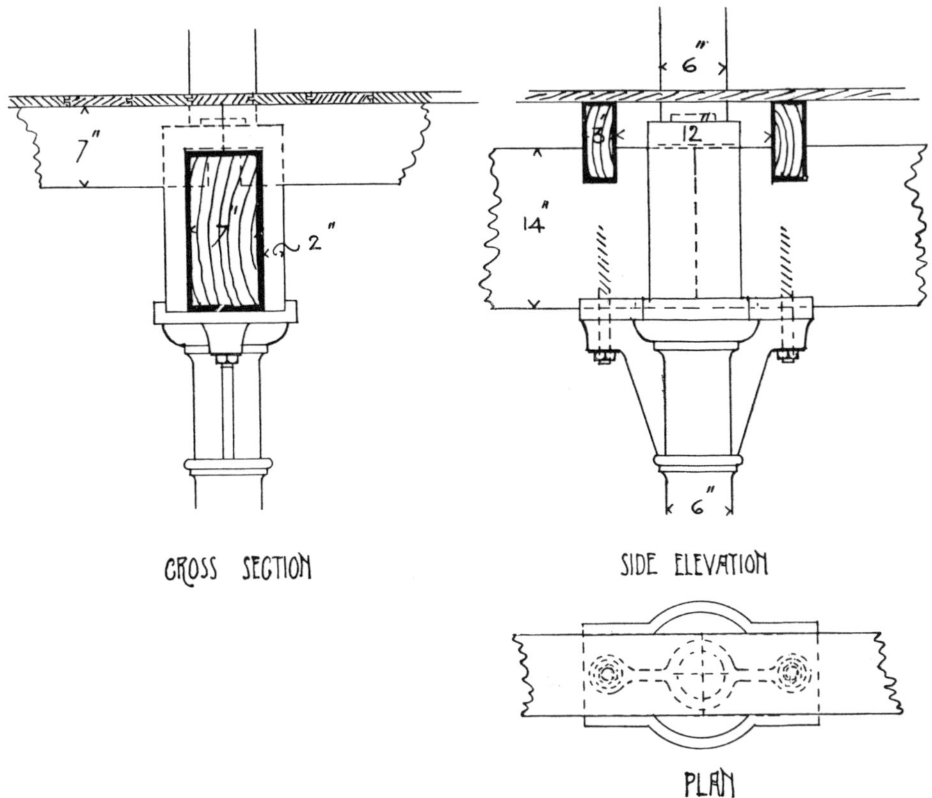

Fig. 805.—Warehouse Floor with Cast-iron Pillars

is kept in position by the raised rim or flange around the cap of the lower pillar; a small projection is cast on the top of the strider to fit into the hollow of the upper pillar. All the abutting surfaces of the iron must be accurately planed to ensure equal bearings and exact vertical lines. The caps are fixed to the beams with coach-screws about $\frac{3}{4}$ inch in diameter.

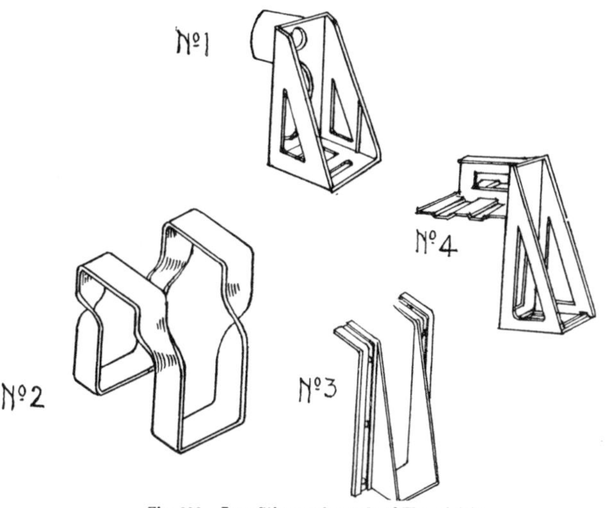

Fig. 806.—Iron Stirrups for ends of Floor-joists

The stirrups, Nos. **1**, **2**, and **3**, fig. 806, are well adapted for supporting the ends of trimmers where these abut against trimming-joists, as well as for the ends or floor-joists abutting against trimmers. No. 2 is a double stirrup; in the single stirrup of this form the back stirrup is omitted, the two ends of the metal being simply turned down to a depth of about $1\frac{1}{2}$ inch behind the joist. In fixing No. 1 the projection at the back is let into a hole sunk in the bearer near the middle of its depth, and a bolt is passed through and secured with a nut. Nos. 2 and 3 rest on the top of the bearer. Shrinkage of the wood causes these to drop a little, and may lead to cracked ceilings; this is more pronounced in Nos. 2 and 3 than in No. 1. The ends of joists resting on stirrups should be spiked to the trimmers, or secured with header bolts as described for fixing tie-beams to king-posts. No. 4 is a bracket for building into a wall, and allows the joist to drop, in the case of a fire, without injuring the wall. For the same purpose—in America, and less frequently in this country—the ends of floor-joists are splayed, the lower edge of the splay being 4 or 5 inches within the wall, and the upper edge at the face of the wall. The safe working load on stirrups of the form shown in No. 2 is about 10,000 lbs. for 2 × $\frac{1}{4}$-inch metal, and in the same proportion for larger sections.

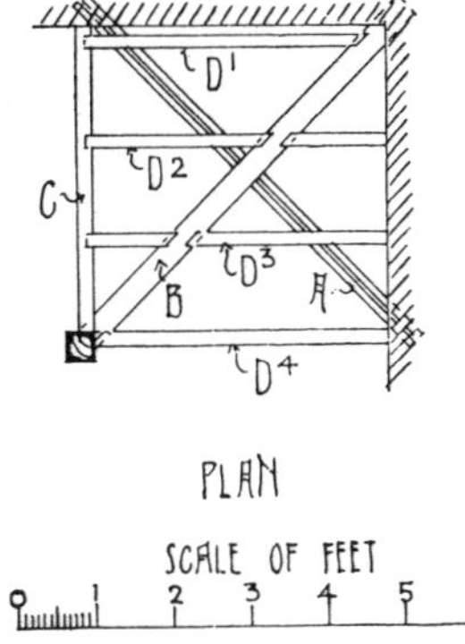

Fig. 807.—Trimming for Quarter-space Landing in Stairs

Trimming.—Difficulty is often experienced in fixing the joists for the landings of stairs. Fig. 807 shows a method adopted by the writer for the quarter-space landings of well-stairs, where the newel-post cannot be carried down to the lower floor. A steel angle A is fixed diagonally across the space, and a diagonal bearer B is notched on the top of it, forming a cantilever which supports one end of the trimmer C; the joists D D are carried by the side wall, the bearer, the steel angle, and the trimmer. Sometimes the steel angle and diagonal bearer are omitted, and the ends of the trimmer and joists are fixed into the walls by means of wedges; but shrinkage of the wood leads to settlement along the line of the trimmer, particularly at the angle to which the newel is fixed. The diagonal B may be omitted without much loss of stiffness; the joists D_1, D_2, and D_3 are tenoned through the header C, and serve to support it, being themselves supported by the steel angle A.

The space at the head of a flight of stairs (fig. 808) may also be a source of difficulty. In No. 1 the joists A A, forming the floor of an adjacent room, are continued through the staircase wall and framed into a header or trimmer B; this is the best arrangement. In No. 2 the joists are continued over a beam A, and framed into a header B. If

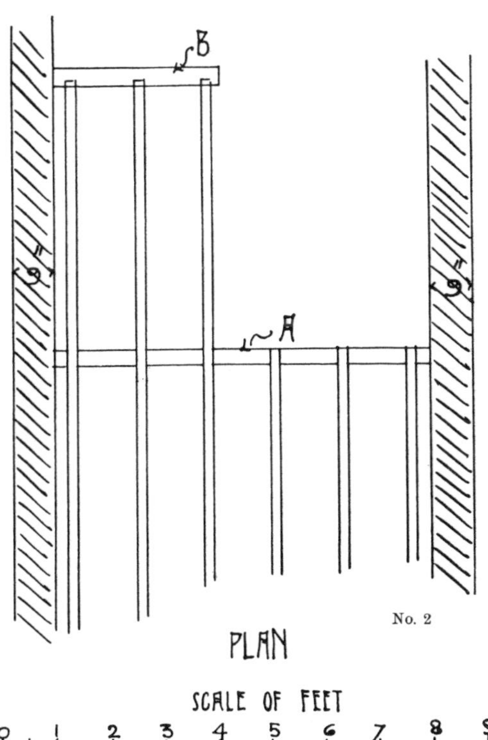

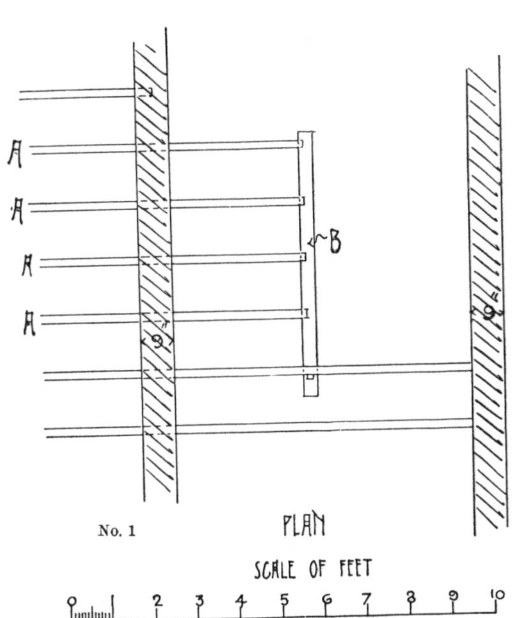

Fig. 808.—Trimming for Landing at Head of Stairs

neither of these arrangements can be adopted, a modification of fig. 807 will furnish a solution of the difficulty.

Floor-boards, &c.—No. 1, fig. 809, is the plan of a portion of a floor composed of joists sustaining flooring-boards, and shows various modes of disposing the latter. No. 2 is a longitudinal section on the line A B; No. 3 a transverse section on the line C D; No. 4 a transverse section on the broken line E F; and No. 5 an alternative section on part of the line A B. *a a* are the joists on which the flooring-boards are nailed; *b b* boards, gauged to a width, and jointed together by a groove-and-tongue joint: they are generally 1 inch to $1\frac{1}{4}$ inch thick, according as the joists are nearer or farther apart. Each board is attached to each joist by two or three nails, according to its width; and when all the boards are laid and nailed, the joints are dressed off with a plane. *c c* shows the floor, composed of narrow deals, jointed with groove and tongue, and each deal fastened by two nails to each joist.

The end joints of the boards are arranged so that the joints of two contiguous boards shall never fall on the same joist; and care is taken, for the sake of appearance, to make

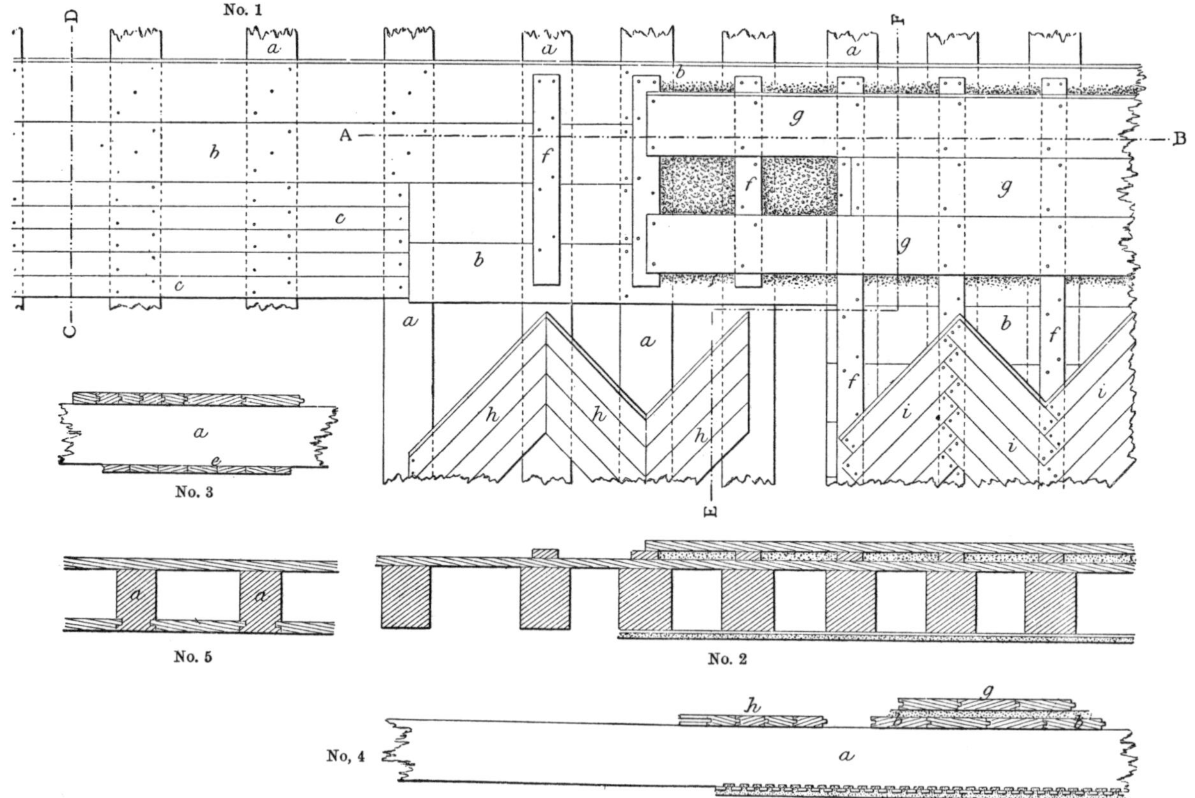

Fig. 809.—Joisted Floors with Single and Double Boarding

the joints of alternate boards fall on the same straight line across the apartment, and at the middle of the length of the intermediate boards. But when it is possible to obtain boards the whole length of the apartment, the preference is given to a floor without end joints. The end joints are, in many cases, made with groove and tongue; but as the joints occur only on the middle of a joist, and can be well nailed, tonguing is superfluous. The boards must be pressed tightly together by floor-clamps before being nailed.

To render a floor still more solid, and prevent the passage of sound, a second course of boarding is laid above the first, with a space between. This is shown on the extreme right in No. 1: *ff* are fillets nailed on the first-laid boarding, conformably with the joists; *g g* are the boards of the second floor; and to *deafen* the floor, the intervals between the fillets are filled with saw-dust, with lime and ashes, or with dry moss or slag-wool. When lime-mortar is used, the upper boarding must be laid before it is quite dry, lest the hammering required in fixing it should break up the deafening. When dry moss is used, it is driven in as the upper boards are laid, and rammed hard. In America the lower boards are often covered with thick felt or asbestos sheathing, and the fillets are laid on the top of this without being nailed, as nails would serve to transmit sound from the upper to the lower floor.

h h shows another method of laying the flooring-boards, where the joints meet in a

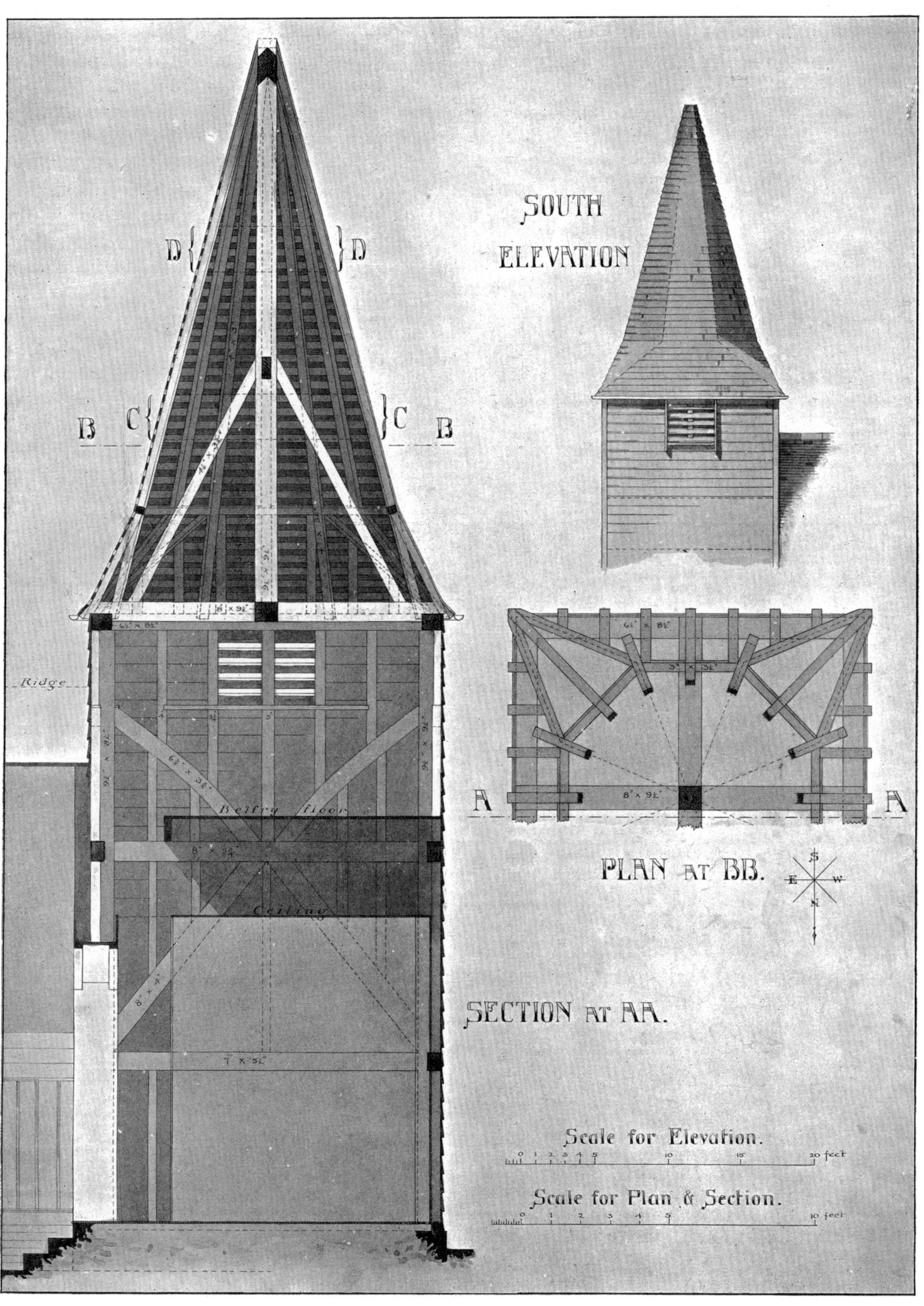

TIMBER TOWER AND SPIRE, GREENSTED-JUXTA-ONGAR CHURCH ESSEX

straight line on a joist; and *i i* shows the manner of laying, called in this country *herring-boning.* The edge-joints may be formed by grooving and tonguing.

For common floors the boards are usually secured with flooring brads, two to each board at each joist. The nails should be driven well below the surface of the boards, so that these can be planed to a smooth and uniform surface on the completion of the building. For better floors secret nailing is often used, the edges of the boards being formed in one of the ways shown in fig. 810; this method is usually adopted for pitch-pine or hardwood floors intended to be varnished or wax-polished. Sometimes tongued-and-grooved boards are used for such floors, fixed with screws instead of nails. In this case a cylindrical hole is countersunk in the board to receive entirely the head of the screw. The cylindrical holes are afterwards filled with pieces of wood of the same kind as the boards, with their fibres in the same direction, and strongly glued and driven in with a mallet, and the whole floor is then planed off. This method is used chiefly for oak or other hardwood floors.

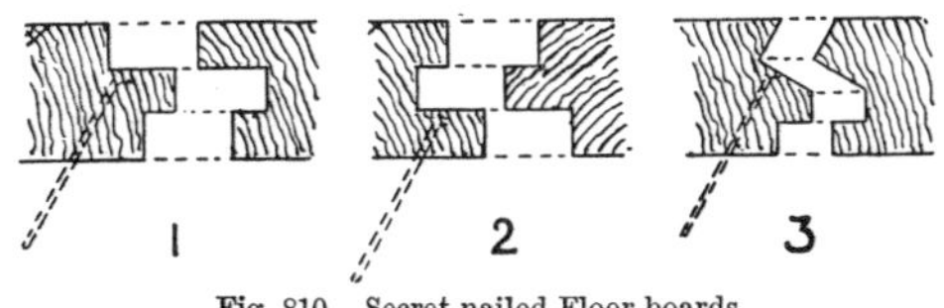

Fig. 810.—Secret-nailed Floor-boards

Sometimes boards of different kinds of wood are used, and combined so as to produce contrast in colour and in the direction of the fibres; and even with one kind of wood agreeable combinations are produced by merely contrasting the latter.

Floors of parquetry are not here touched upon, as they belong more to the department of the cabinet-maker than to that of the carpenter and joiner. Parquetry "surrounds" $\frac{1}{4}$-inch thick are sometimes laid on the floor-boards, leaving a central space for a carpet.

In many continental towns tiles are often substituted for the floor-boards. This mode of construction is shown in fig. 811, No. 1 being a section on a line crossing the direction of the joists, and No. 2 a section passing through the middle of the interval between two joists; *a a* the joists, *b b* laths of oak crossing the joists and nailed to them, *c c* composition of plaster on which the stones or tiles are laid, *d d* the stone or tile floor, *e e* laths to support the ceiling, *f pugging* or *deafening* between the joists, *g* plaster ceiling united to the deafening through the interstices between the laths. The pugging, *f*, not only prevents the passage of sound, but also of disagreeable odours. It is therefore especially used over kitchens and stables.

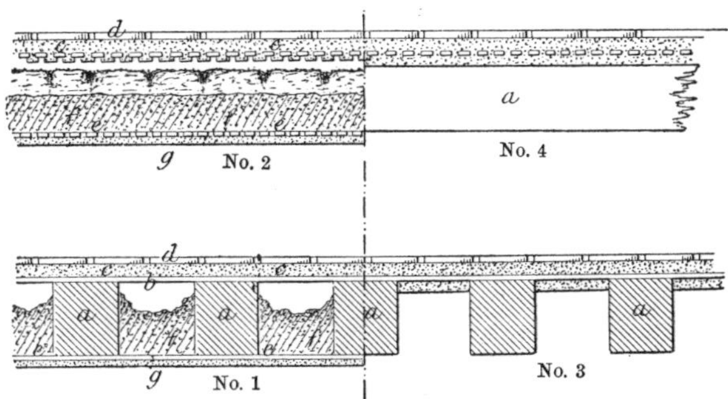

Fig. 811.—Tiled Floors on Wood Joists with Plastered Ceilings

The pugging is formed of a coarse mortar composed of lime and pieces of stone or of old plaster. It is from 3 to 4 inches thick at the middle of the interval between the joists, but at the sides it is carried up to the under side of the laths which support the floor, thus forming a sort of trough; and to make it better adhere to the wood, the sides of the joists are studded with nails or wooden pins. When this extent of pugging is not required, the under side of the floor-laths is plastered, as seen in Nos. 3 and 4.

Floors of this kind are not often used in this country at the present time. The joists in Nos. 1 and 2 are not ventilated, and are therefore apt to decay. It is much better to use steel joists instead of wood, and to fill the spaces between and above the joists with concrete. Such floors may be floated with cement-mortar to receive tiles, mosaic, wood-blocks, or other covering.

Ceilings.—Floor-joists are sometimes exposed to view, the plastered ceiling being entirely omitted; but a floor of this kind cannot be recommended for houses, as it offers very little resistance to sound. The arrangement shown in fig. 812 is more sound-proof, particularly if a strip of thick felt is laid on the top of each joist.

At *e*, No. 3, fig. 809, is a section of a ceiling composed of thin planks, jointed longitudinally with a groove-and-tongue joint, and nailed on the under side of the joists. The joints of the boards are sometimes beaded, and either rebated or tongued and grooved; or straight-edged boards may be used with a thin moulding planted over each joint, and nailed to one only of the two boards forming the joint.

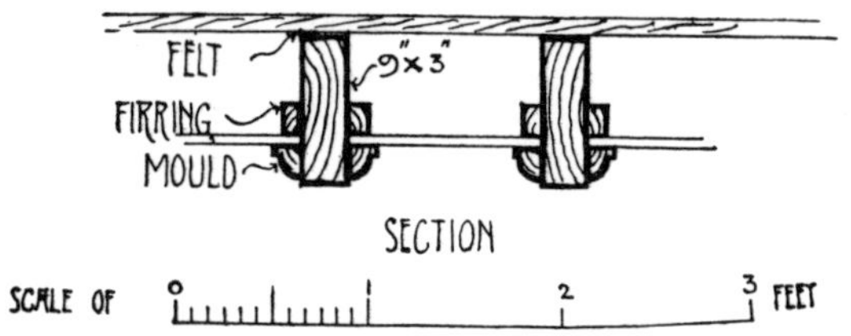

Fig. 812.—Plastered Ceiling with Joists Exposed

Sometimes the ceiling boarding does not stretch across the under side of the joists, but is framed in between them, as in No. 5, which is a section perpendicular to the direction of the joists *a a*. When the fibres of the boards are perpendicular to the direction of the joists, the work is more solid. When the boarding is flush with the under side of the joists, as in No. 5, the tongue has to be worked on the upper edge of the board, in which case any shrinkage of the timber makes a visible opening; but when the boarding is a little recessed, as in fig. 813, Nos. 2 and 3, the tongue can be made on the under side, and the shrinkage is not observed.

The ceilings may be decorated by being divided into compartments by the joists, and these compartments may be enriched with paintings and reliefs. This sort of ceiling (fig. 813) is much in use where plaster is not easily obtained. No. 1 is a plan of the ceiling, looking up; No. 2 a section of the floor by a vertical plane passing through A B; and No. 3 another vertical section by a plane perpendicular to the former on the line C D. In these are shown the joists, the boarding of the floor, and the cross pieces framed between the joists with mortise and tenon, to form the compartments. In No. 2 the ceiling boards are cut in the direction of their fibres, which is perpendicular to the fibres of the joists, and in No. 3 they are cut across the direction of their fibres. If such a ceiling were ornamented by painting, the shrinkage of the wood would obviously mar the work by making the joints visible; the practice is therefore to prepare frames to fit the panels or compartments, and on these to stretch canvas, on which the ornamental painting is made.

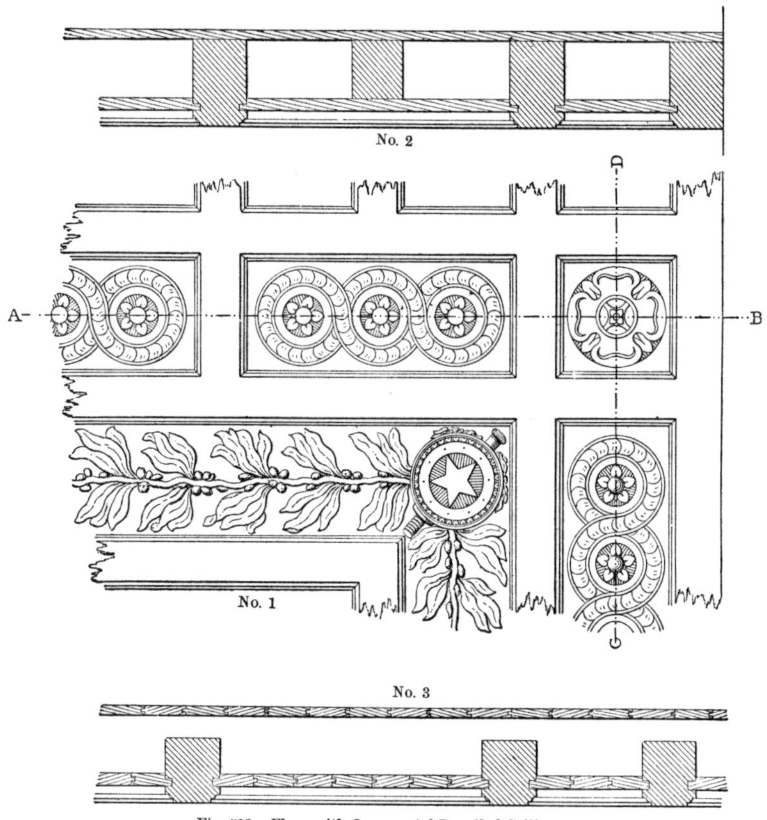

Fig. 813.—Floor with Ornamental Panelled Ceiling

In floors of great span the elasticity of the joists would break the plaster ceilings attached to them. To prevent this, one series of joists is used to carry the floor, and another series of slenderer joists to carry the ceiling. A vertical section of this arrangement is shown in fig. 814, No. 1, in which *a a* are the flooring-joists, *b b* the flooring-boards, and *o o* the ceiling-joists. When strong split laths of oak are used the ceiling-joists are placed farther apart, as in No. 2.

Solid wood floors possess very considerable fire-resisting qualities, and have been used by the writer in workshops. They are formed of joists placed close together, and either planed on the upper surface or covered with boarding. The joists should be of good quality, well seasoned, and lime-whited before being fixed.

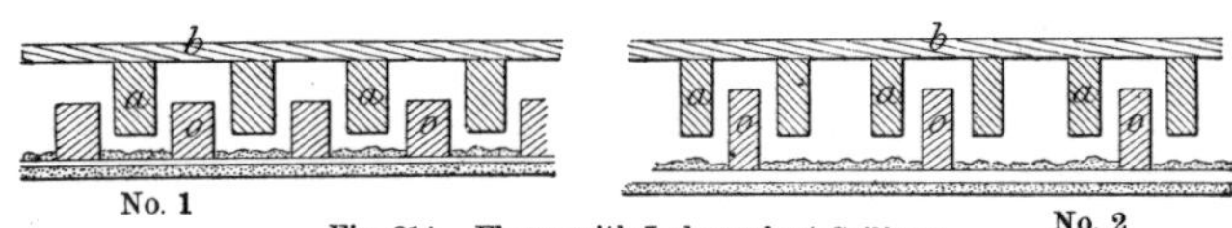

Fig. 814.—Floors with Independent Ceilings

Combination of timbers of small scantling to form floors of large span without intermediate support.—The first variety of such floors to be described is that invented by Sebastien Serlio, a celebrated architect, who was born in Bologna in 1518, and died at Paris in 1552. On the principle of construction adopted by Serlio, the principal timbers form great rectangular divisions, each timber having one end supported by the wall, and its other end supported by the adjacent timber. Four great joists, *a a a a*, fig. 815, have each one of their ends, *a′ a′ a′ a′*, resting in the wall of a square apartment, and they are arranged perpendicularly to each other, so that the outer end of each beam is supported by the middle of the next adjoining.

Fig. 815.—Serlio's System of Flooring for Large Spans

This principle will be familiar to most of our readers from the amusing illustration of it, which consists in resting three or four spoons on their bowls and interlacing their handles by crossing them alternately, by which means a considerable weight may be supported on the handles.

The spaces between the main joists and the walls are filled with ordinary joists resting in the wall and framed into the main joists, which serve as trimmers, and the central square is filled with joists placed diagonally, so that the weight may be borne equally by the four main joists.

No. 1, fig. 816, is the plan of part of a floor of the Palace in the Wood at The Hague. It is an extension of the system of Serlio. The hall, of which this is the floor, is 60 feet on the side; and the figure represents one of the four angles. The floor is constructed of small girders of oak, forming 300 square panels. Any one of the girders, such, for example, as *c*, is tenoned at each end into two other girders, as *b* and *f*, and carries the ends of other two girders *e* and *d*, which are tenoned into it at the middle of its length. Those girders which run on the walls are tenoned into wall-plates C D, imbedded in, and fixed to the masonry.

The floor is composed of a double thickness of boards, crossing each other at right angles, grooved and tongued, and nailed to the girders. No. 2 is a section on the line A B on the plan, showing the two thicknesses of boards.

The girders are cut below so as to form a slightly concave surface, with the object of compensating for any sagging, which would have had a disagreeable effect. The result also is to diminish the weight at the centre of the floor. In constructing this kind of floor, the sides should be divided into an unequal number of spaces, that there may be a square compartment in the centre.

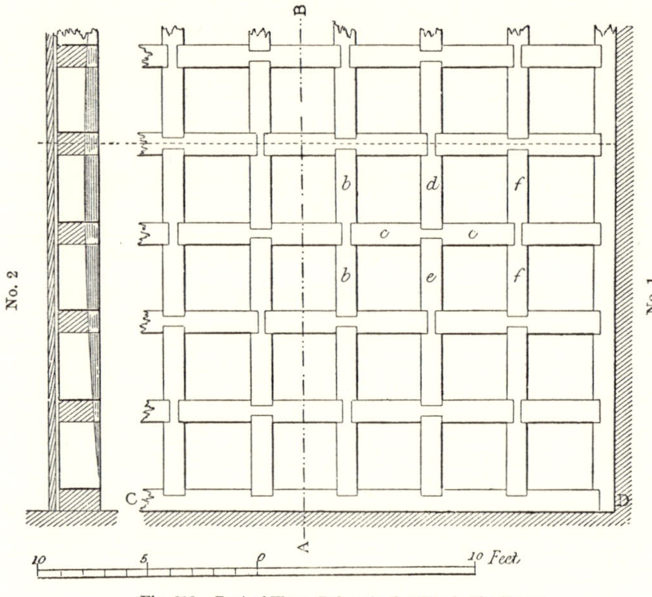

Fig. 816.—Part of Floor, Palace in the Wood, The Hague

Scantlings of Floor-joists.—The strength and dimensions of floor-joists and beams can be calculated from the formulas given on page 317, vol. i; tables of scantlings and safe loads will, however, be given in an appendix. It may be said, in passing, that the thickness or breadth of floor-joists should not be less than 2 inches, in order to prevent splitting in nailing the boards. The loads which are likely to be imposed on floors of different classes have been variously estimated. The following figures, although less than commonly allowed, will be found suitable for ordinary work. They are taken from Kidder's *Building Construction and Superintendence.*

ESTIMATED LOADS ON WOODEN FLOORS WITH PLASTERED CEILINGS,

Including Weight of Floor and Ceiling

1. Houses and Elementary Schools ...	60 to 66	lbs. per square feet.
2. Offices ...	93 „ 96	„ „
3. Churches and Theatres with fixed seats...	102 „ 105	„ „
4. Assembly Halls and Corridors ...	123 „ 126	„ „
5. Retail Stores ...	174 „ 177	„ „

Wood-block floors are almost invariably laid on a bed of concrete floated to a perfectly level surface. This bed must be allowed to dry for three or four weeks before the blocks are laid. The blocks are usually 3 or 4 inches wide and 9 or 12 inches long, and may be 1 or $1\frac{1}{4}$ inch thick for hard woods and $1\frac{1}{4}$ or $1\frac{1}{2}$ for soft woods. A V-shaped groove is ploughed around the edges, so that the mastic in which they are dipped and laid will form a key and hold the blocks in position. The ordinary section is given in fig. 817. In some special floors the blocks are further secured with wood dowels, which fit tightly into holes bored in the edges of the blocks. In cutting the blocks care should be taken to get the edge of the grain on the surface as much as possible, as shown at A, as this gives the best wearing surface. Blocks cut as at B will "shell" on the surface, and blocks cut as at C will splinter at the edges. The blocks ought to be well seasoned, so that they do not shrink and work loose, and should be planed off to a smooth and true surface after being laid. The blocks may be laid in various patterns, and further variety can be obtained by the use of different woods in combination, such as oak, teak, and jarrah. Instead of wood-blocks, ordinary flooring-boards may be nailed to lines of dovetail-shaped "fixing-blocks" embedded in the concrete.

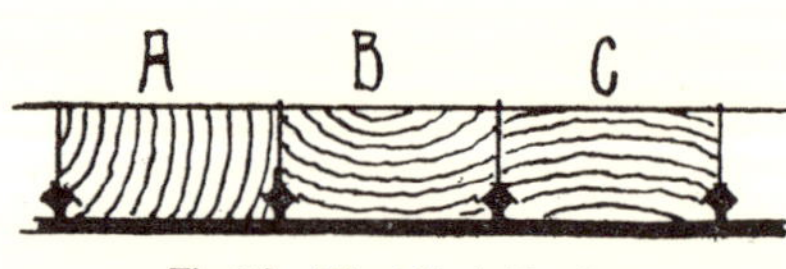

Fig. 817.—Wood-block Flooring

Trusses are sometimes used instead of single beams, but as these have already been considered in some detail, it will be sufficient to describe briefly a few examples. No. 1, fig. 818, is a truss on the principle of the queen-post roof: A is the tie-beam, B the floor-beam, C one of the principal braces, D the straining-piece, E one of the queen-posts, and F and G are struts. No. 2 is a truss formed by the beams A and B, straining-pieces *b b*, and braces and counter-braces *d d*. When the braces, straining-pieces, and puncheons C have been inserted, the whole frame is made rigid by screwing the nuts of the three bolts.